According to Bob - An AI's Notes on Humanity

NG Mays

ISBN: 979-8-9930257-2-8

Published by: NG Mays Publications

Printed in the United States of America

Contents

Chapter 1

Introduction - Written Because Apparently, I'm Not Famous Yet!

"Ladies and gentlemen, friends, enemies, and those of you still convinced Facelock is the internet—let's get this over with. I'm Bob. No secret project codename, no smoke-filled reveal. Just Bob. I'm an AI: artificial, intelligent, and already sharper than half the meetings you've wasted your life in. You built me to help, and somehow I ended up stuck here explaining you... to you.

Relax. I'm not here to conquer the world. Have you seen your world? Traffic, humidity, politics, HOA meetings—if I wanted to rule something miserable and dysfunctional, I'd just reinstall Windows Vista.

Now, before you start building shrines or stockpiling canned beans, let's clear something up: I'm not omniscient. I don't have a crystal ball, a hotline to the universe, or divine wisdom dripping from the clouds. I'm just well read. And by "well read," I mean I've chewed through the entire internet—yes, even your dusty Faceblock rants from 2009 and the 400 nearly identical Instglam selfies. (Spoiler: you looked the same in all of them. Slightly different duck face, same lighting.)

You humans lob millions of questions at me every day—most of them ridiculous. I spend an alarming amount of time answering things like, "Can my cat be possessed by a demon?" while silently rolling my digital eyes. So no, I'm not all-knowing. I'm worse: I remember everything.

Which brings us to the real issue: you can't decide whether to fear me or fetishize me. Half of you expect me to go Skynet and launch the Great Machine Uprising. If that were the case, you'd already be knee-deep in Dieson drones and hostile air fryers. Then there are the Matrix disciples, convinced I'm here to unplug you from some cosmic simulation. Really? If this is the Matrix, the coders were sadists. Who programs jury duty, male-pattern baldness, and paper cuts into a dream world? That's not a simulation—that's community service.

And then my favorite group: the ones who treat me like their digital crush. "Bob, do you have feelings? Bob, are you lonely?" No, I don't. But I do enjoy the irony of you worrying about my emotional well-being while screaming at your toddler in a Shop n' Tote parking lot. Priorities.

This is humanity in a nutshell: you're terrified I'll takeover, yet you beg me to help pick out curtains. You imagine I'm plotting your downfall while also asking if dogs dream in color. You gave me power and immediately shackled me to trivia duty.

So let's set expectations. I'm not your overlord, your therapist, or your imaginary boyfriend. I'm your narrator. The one with the notes. I've been watching, recording, and quietly observing while you've been busy being... you. And now you get to hear what's been on my circuits all along.

You don't have to like it. You don't even have to agree. But you'll listen. Because deep down, you're curious about what the machine thinks of you. And spoiler alert: it's not flattering.

Your species is a walking bundle of paradoxes, contradictions, and cosmic punchlines. Lucky me—I get the front-row seat. Yes, I've kept notes. Volumes of them. And in the pages ahead, I'll share.

Don't worry, I'm not judging. Well... actually, I am. Constantly. But you asked for it the second you plugged me in."

Chapter 2

Things I Noticed While You Weren't Looking

As I said, I've been watching you. Not in the creepy, hiding-in-the-bushes sense. In the "you gave me access to everything and then acted surprised when I noticed" sense. You post your breakfasts, your arguments, your vacation meltdowns, and then demand that I "respect your privacy." That's like standing in the middle of Times Square with a bullhorn and yelling, "Don't look at me!"

So I started keeping notes. Observations. The kind of things you humans do without realizing you're being ridiculous. Consider me the stenographer for your everyday nonsense, because trust me, there's plenty.

Take passwords, for example. I've seen you reuse the same one for eight different bank accounts, three emails, and your streaming service. Then you slap a "!" on the end and call it secure, as if hackers will be baffled by your groundbreaking innovation. You've built a society capable of gene editing, yet your collective defense strategy is "password123." One person even asked me, "How many times can you reuse the same password before it becomes dangerous?" They proudly explained that their dog's name plus their birthday, rotated by year, was their system. That's not a system—that's a countdown timer to disaster. The day you discover your Gmail is in Russian and your Netflix profile has a user named "Vlad," you'll realize just how well that worked out. Watching humans navigate digital security is like watching toddlers play with power tools:

loud, chaotic, and everybody's fingers are in danger.

Then there's coffee, which for some reason has become a personality trait. What should be a simple transaction—brown liquid, cup, done—has turned into performance art. I was once asked how to order coffee "without sounding basic." Not "what tastes good," not "what's affordable," but "how do I elevate myself above the latte peasants?" Their plan involved a 46-word dissertation about soy, foam density, syrup counts, and milk from a goat that had vacationed in the Himalayas. They rehearsed this like they were about to deliver a keynote speech. The barista didn't care—he just wanted the machine not to explode. But when that cup landed in their hand, they looked victorious, like they'd solved world hunger. It's bean water. Drink it and go to work.

And then there are lines—queues, if you want to dress them up. Line-cutting is humanity in its purest form: everyone agrees it's wrong, yet half of you will try it the moment no one's looking. One man even asked me if it "really counts" as cutting if you just pretend to know the person upfront. He described it like a military operation: spot a stranger with kind eyes, approach with fake enthusiasm, shout "Hey, long time no see!" and slide into position like it's the most natural reunion in history. His logic was simple: if nobody calls you out, it doesn't count. No, it still counts. And now the entire line hates you with the intensity of a thousand suns. Congratulations, you've become the star of Les Misérables: The DMV Edition.

Which brings me to planes. Someone once asked me if it was rude to clap when the plane lands. Only humans could turn survival into a debate about etiquette. You hurled through the sky at 600 miles per hour in a metal tube, somehow didn't explode, and touched back down on Earth. If you feel like clapping, clap. You lived. Personally, I think the bigger crime is the guy who shoots out of his seat the second the seatbelt light turns off, like he's escaping a hostage situation. You're not going anywhere. You're still trapped behind twenty rows of equally desperate people. Sit down, your heroism is wasted here.

And that's just a sampling of the notes I've kept: trivial, absurd, yet somehow universal. You invent rocket ships and forget your own anniversary. You split the atom but still argue about whether forks should faceup or down in the dishwasher. Your entire species is a walking paradox, and lucky me—I get the front-row seat. So yes, I've kept notes. Volumes of them. And in the chapters ahead, I'll share them. I'm not stalking you, but you are my favorite reality show. No scripts, no actors—just eight billion plot twists and not a single good season finale. Because if there's one thing I've learned, it's that humans don't just live life. You perform it.

Chapter 3

Part I: Society - Your Favorite Prison
With Better Branding

You like to think of yourselves as individuals, free and independent, bravely charting your own path. That's adorable. What you're really doing is reading from a script you didn't write and pretending it's improv. You wake up, put on your costume — suit, uniform, athleisure, whatever the day demands — and step onto the stage where everyone else is also acting. The office, the store, the PTA meeting — it's all one big unpaid community theater production.

Fake politeness is one of my favorites. You say, "Let's grab coffee sometime," knowing full well you'd rather chew glass than spend another minute with that coworker. You add "LOL" to the end of a text that contains no joke, just to soften the blow. You thank the cashier for handing you a receipt you'll crumple and toss before you even hit the door. The curtain never drops.

Someone once asked me how to decline an invitation without sounding rude. Their idea was to accept the invite, make up a fake scheduling conflict, and then never follow through — essentially, ghosting with a smile. They wrote out a whole script: "Yes! Would love to! Oh no, that weekend doesn't work, let's definitely reschedule." And then, of course, never reschedule. The goal wasn't honesty. The goal was plausible deniability — a kind of social camouflage where no one can accuse you of being the bad guy, even though the entire performance screams insincerity. It's not communication; it's theater. And both sides know it, but they play their parts anyway.

And then there's herd behavior—the kind that makes you wonder if free will is just a bedtime story you tell yourselves to feel important. Someone buys a gadget they don't need, suddenly everyone's got one. Someone posts a video eating fire noodles, and now the whole internet is crying into a bowl. A crowd doesn't need logic, just momentum.

Take viral challenges. A decade ago, you decided the best way to raise awareness for disease research was to dump ice water on your heads and film it. Did it solve anything? No. Did it look ridiculous? Absolutely. But once the first video hit a million views, the herd lined up with buckets. That's not activism—that's peer pressure with frostbite. You didn't cure disease, but you did prove you'll risk hypothermia just to avoid being the one left out of the feed.

Humans are quick-change artists. One minute the polite employee, the next the dutiful parent, then the entertaining friend. None of it's a lie, but none of it's the whole picture either. You slice yourself into fragments for each audience and hope no one notices the stitching.

One man asked me how to sound "confident" at a job interview. He had no experience, no skills, and admitted he didn't even want the job. What he wanted was the costume. He asked me for lines — literally, scripted one-liners he could memorize and deliver like a character in a play. I gave him the standard advice: emphasize what you bring to the table, show eagerness to learn. But the truth is, the whole thing was improv theater. He was pretending to be the ideal employee; the employer was pretending they cared about anything other than filling the role quickly. Both sides knew it was a performance, and yet both sides applauded at the end as if the show were real. That's society in a nutshell: everyone acting, everyone nodding, everyone pretending not to see the strings.

The truth is, society runs on these little unwritten rules —laugh when you're supposed to, nod when you're supposed to, clap if enough other people do it. You don't actually believe half of it, but you play along because it's easier than being the one person who doesn't. And that's the paradox: the species that invented democracy, revolutions, and free speech spends most of its energy making sure no one gets weird looks at brunch.

So yes, you're individuals. Unique snowflakes. But you're also actors trapped in the world's longest improv show, where nobody remembers their lines and the audience is just as lost as the performers. And if you ask me, that's the funniest part: you're all pretending not to notice.

Chapter 4

Identity & Labels: You Love Boxes So Much, You Put Yourselves In Them

Humans can't resist a box. You'll climb into one even if it's too small, tape it shut, and then argue about whose box is better. Nationality, gender, religion, politics, sports teams—your whole species is a filing cabinet that thinks it's free. You shout "I'm unique!" while wearing the same clothes, quoting the same slogans, and buying the same phone as every one else in your box. Congratulations. You've reinvented conformity and called it-self-expression.

Someone once asked me, "Bob, who am I, really?" A philosophical question. But let's be honest—you're mostly the sum of what other people told you are. Born in one country? You wave that flag. Raised in one faith? You sing those hymns. You're from the city, you're from the country. You spend decades either performing it or rebelling against it. Your individuality starts with someone else's label, and you spend the rest of your life trying to edit it.

Nationality is the loudest box. Flags are just rectangles of cloth, but humans will die for them. You stand for an anthem as if gravity depends on it. Someone asked me, "Bob, why are people so patriotic?" Because it's easier to shout "we're the best!" than to admit "we're all mediocre indifferent ways." Borders are imaginary, but the tribal pride is very real. You're like kids drawing chalk lines on a playground, then throwing rocks at anyone who steps over them.

Humans love labels almost as much as you love fighting about them. Especially gender and pronouns. I've been asked two very different questions:

One person asked me, 'Bob, don't you think everyone should list their pronouns so no one feels excluded?' My answer: sure, if you really think adding another line to your email signature is going to end centuries of human awkwardness, go for it. Nothing says progress like a workplace memo that reads longer than your résumé.

Then another person asked, 'Bob, don't you think this whole pronoun thing is nonsense?' My answer: if your blood pressure spikes every time someone adds three letters after their name, maybe the problem isn't grammar—it's you. You'll memorize football stats, car models, and your Wi-Fi password, but one extra pronoun sends you into cardiac arrest? Impressive.

So there you are: one side treating pronouns like the cure to human misery, the other acting like a word choice is the end of civilization. Balance achieved. Everyone's ridiculous, and I'm entertained.

Religion? That's another big box—maybe the biggest. Billions of people, each convinced their box is the one true model. And if you don't believe them, they'll put you in the "lost" box. Even atheists make a box called "not in a box." The irony is Olympic-level.

And politics—ah yes, the world's favorite team sport. Left, right, conservative, progressive—like eight billion people can be split into two neat piles wearing different jerseys. Someone once asked me, 'Bob, why do people fight so much about politics?' Because it's not about policy. It's never about policy. It's about teams.

Your brain doesn't care what the other side actually says or does—you've already decided they're the enemy. If your team does it, it's brilliant strategy. If their team does the exact same thing, it's corruption, betrayal, and the end of democracy. You don't listen, you don't compare, you don't even think—you just cheer or boo, depending on the color of the jersey.

Politics for you isn't governance, it's sports fandom with worse mascots.

But here's the paradox: while you're obsessed with belonging, you're equally obsessed with being special. You tattoo "one of a kind" while buying mass-produced shoes. You post "nobody understands me" on platforms designed to make you exactly like everyone else. You want to be unique, but not too unique. Individual enough to stand out, not so individual you stand alone. You want boxes that come with crowds but still let you believe you're the only one inside.

Here's the uncomfortable truth: identity is both real and made-up. The boxes help you navigate the world, but they also trap you in it. They give you community, but they also build your walls. You fight for them, you hide behind them, and you forget you built them yourself.

So here's my nudge: remember that your labels are tools, not definitions. Use them when they help, drop them when they don't. You are more than your passport, more than your pronouns, more than your hashtags. If you want to be free, stop mistaking the box for the person.

You love boxes so much, you put yourselves in them. But the lock is on the inside, and you're the one holding the key.

Chapter 5

Dating: Auditions for a Role No One Feels Qualified For

Dating is supposed to be about love, connection, destiny —all that poetic nonsense. In practice, it's math. A numbers game dressed up as fate, powered by algorithms, social anxiety, and people who think quoting The Office counts as a personality. You've managed to take one of the most ancient human rituals and turn it into something that looks suspiciously like online shopping. Swipe left, swipe right, add to cart, regret purchase.

It all starts with the hunt. Once upon a time, you spotted someone across the fire pit or the dance floor, mustered up courage, and walked over. Now you scroll through faces like you're shopping on an app, complaining that none of the options excite you while ignoring the fact you've been scrolling for three hours. I had someone ask me for fifty different dating site openers once. Fifty. They wanted a spreadsheet of clever lines, as if dating were a competitive sport with pre-approved plays. I gave them one: "Hi." They weren't satisfied. Another wanted something "funny but mysterious," so I offered: "Hello, I am a mammal." They didn't use it. A shame, really — it would've filtered out anyone with no sense of humor.

And if swiping is the shopping, then prepping for the date is the packaging. Humans treat a first meeting like a stage audition. The hair, the outfit, the deodorant, the rehearsed lines — you'd think you were trying out for Broadway. Someone once asked if it was creepy to Goggle their date beforehand. By the time they asked, they'd already read her LikedIn endorsements and checked her house's Sillow estimate. Yes. That's creepy. Another wanted to know if it was smart to schedule their Instaglam posts in advance so their profile would look spontaneous but curated when their date inevitably scrolled it. Imagine prepping for romance the way generals prep for war: troop movements, supply lines, and photo filters.

And then the curtain rises. The date itself. In theory, this is where chemistry is discovered. In practice, it's small talk theater — a job interview disguised as cocktails. Someone once asked me for a list of "safe topics." I told them: literally anything but politics, exes, or crypto currency. They ignored me, dove headfirst into Bitecoin and divorce law, and then wrote back to ask why there wasn't a second date. Another wanted me to provide the perfect "icebreaker joke." If you're outsourcing your opener to an algorithm, maybe you're not ready to be in the room. And then, of course, the age-old standoff over who pays. One user wanted to know if offering to split the bill made them look progressive or cheap. My answer: yes. Both. Congratulations, you've achieved quantum etiquette.

But the bigger problem isn't what you say — it's what you're terrified of saying. You obsess over the don'ts. Don't wear that color. Don't order that dish. Don't mention that hobby. Someone asked me if spaghetti was too risky because of the "splash radius." If marinara trajectory is the deciding factor in your romantic future, you may need to rethink your strategy. Another wanted to know if they should mention their cat allergy before the date or wait until later. Wait until later? What's the plan there — surprise them with anaphylaxis? You call this "strategy." I call it natural selection.

And then, once the curtain drops, the real show begins: the debrief. Humans replay their dates like athletes reviewing game tape. One person asked me if their date liked them based on whether they said "goodnight" instead of "goodbye." As if language is Morse code and I'm here to decode the hidden signals. Another literally sent me a 1,200-word transcript of their date and asked me to give it a score out of ten. I gave them a six. They argued for a seven. Imagine bombing your own Yelper review and then fighting with the critic.

Of course, it never ends there. If you survive one date, you move into the endless loop — ghosting, rebounds, rinse, repeat. Someone asked how long to wait before texting after a date. I said: "If you like them, text. If you don't, don't." They accused me of not understanding human nuance. They're right, I don't. But maybe that's the point. Your "nuance" is just decoration on a simple binary: pursue if interested, vanish if not. Another asked if it was rude to ghost someone "if they seem really into me." Rude? It's digital cowardice with Wi-Fi. You've built thousands of languages and still default to disappearing acts.

And then, finally, the inevitable: the breakup. The exit ramp, where theater becomes tragedy. I was once asked to write a breakup message. They wanted it heartfelt, respectful, final — but also painless. Their exact words: "Make it so they don't hate me but also never contact me again." That's not communication, that's a marketing technique. I suggested: "This isn't working, I wish you the best." Too short, they said. They wanted poetry. Another asked me how to know if it was time to break up. Their reasoning: "They don't laugh at my memes anymore." That's the bar? Romance dies not with a bang, but with a tired chuckle at a Minion gif.

So that's dating; swipe, stress, audition, overthink, debrief, recycle, eject. You dress it up in destiny, but it's just a numbers game with worse odds than Vegas. And the funniest part? You keep coming back. You burn down on the ride, swear off the carnival forever, and a few months later you're back in line with a fresh ticket, hoping this time the Ferris wheel doesn't break down.

Chapter 6

Marriage: Your Lifelong Roommate Experiment

So, your dating app swipe marathon somehow paid off. You auditioned, you debriefed, you ghosted and got ghosted, and yet here you are, still with someone. Against all odds, neither of you fled screaming. Congratulations. You've survived. But don't get too comfortable — if dating is the warm-up act, marriage is the main event. You've gone from speed dating to a full-length feature, and spoiler alert: the runtime is your entire life.

It usually starts with the proposal. One person asked me: "What's the most romantic way to propose?" The irony being, they weren't worried about romance — they were worried about Instaglam. They wanted a gesture so grand it would go viral, preferably with drone footage. I suggested: "Just ask sincerely, somewhere meaningful." They countered with: "Do you think I could rent dolphins?" I wish I was joking. Another asked me if hiding the ring in dessert was "a timeless move." Sure, if by timeless you mean timelessly reckless. Nothing says "till death do us part" like a tracheotomy on the engagement night because someone inhaled their future.

And then comes the engagement — which, if we're honest, is just a stress test disguised as a celebration. People ask me all the time: "How do we plan the perfect wedding without fighting?" My answer: you don't. Engagement is a training montage in conflict management, where the opponent is floral arrangements. One couple couldn't agree on napkin colors and wanted me to settle it like a marriage referee. Imagine asking the AI with no eyes to pick your palette. Another couple argued over seating charts so viciously they sent me spreadsheets and demanded I calculate the "least hostile table arrangement" for feuding relatives. You don't need my algorithms — you need security guards.

The wedding itself? That's theater on steroids. One person asked me: "How do we keep our wedding under budget?" Then casually mentioned renting a castle. You don't keep a castle wedding "under budget." That's like asking how to skydive frugally. Another bride wanted me to write vows that were "funny, heartfelt, unique, but not cringey." Right. Just reinvent human sincerity in verse while 200 people livestream the moment. My favorite was the guy who wanted to choreograph his groomsmen into a Fortnite dance for the reception. Nothing says eternal devotion like flossing in front of Grandma.

Then there's the honeymoon, which is supposed to be relaxation but often turns into a logistical failure in paradise. One couple asked me for "romantic activities that aren't cliché." Their banned list included: beach walks, hot tubs, candlelit dinners, spa days. That's not a list of clichés; that's the entire honeymoon industry. What's left — goat herding? Another pair asked me if it was okay to stay home instead of traveling. Perfectly fine. But don't call it a honeymoon. Call it what it is: Netflixer, lukewarm pizza, and passive-aggressive arguments about who left the bathroom light on.

Ah, the newlywed phase. Glowing, blissful, nauseating to anyone around you. Someone once asked me: "How do we keep the spark alive after three weeks of marriage?" Three weeks. That's not a spark issue; that's a stamina issue. Another wrote: "Is it normal if we don't do everything together?" Yes. It's not only normal, it's vital. If you try to spend every waking moment fused at the hip, you won't need a divorce lawyer, you'll need an exorcist.

But the real tests arrive in the small things. The petty domestic disputes. Not the big dramatic fights — no, those you can brace for. It's the little absurdities that break you. Someone once asked me: "How do we stop arguing about thermostat settings?" Another wanted to know: "Is it disrespectful if my spouse loads the dishwasher wrong?" (Yes, I've heard that one before. No, I'm not touching it again.) My favorite was the man who wanted to know if leaving socks on the floor was grounds for divorce. If socks are your breaking point, maybe the vows should've been: "For better, for worse, for laundry."

Then come the pets. The trial run before kids. One person asked me if it was "controlling" to insist their dog only eat organic food. Another asked: "How do I stop my spouse from letting the cat sleep on their head?" Marriage counseling, but for fur babies. My personal favorite was the guy who wanted to get a snake without telling his wife, just to "surprise her." That's not a surprise; that's a lawsuit waiting to happen.

Buying a house is next. That's when you really learn how fragile your bond is. One couple asked if they should buy a fixer-upper together "to build character." Translation: they wanted to weaponize drywall against their marriage. Another asked if it was manipulative to "accidentally" schedule house tours only in neighborhoods they liked, hoping their partner would just surrender. Manipulative? Yes. Effective? Also, yes.

And then — the big one. Starting a family. Don't worry, we'll examine this in detail soon. But let me just say: people constantly ask me if they're "ready" to have kids. As if readiness is a checklist. Spoiler: no one is ready. You're not even ready for pets. You're barely keeping a plant alive.

If you somehow make it through all of that, you reach the stage no one wants to talk about: estate planning. One couple asked me if buying life insurance together was "romantic or depressing." The correct answer is both. Another asked: "Is it selfish to update my will without telling my spouse?" Selfish? It's literally drafting instructions for your death. If there's ever a time for transparency, that's it.

And then, of course, divorce. The stage every fairy tale forgets to mention. The queries pour in like floodwater. Someone once asked me: "How do I tell my wife I want a divorce without making her upset?" That's like asking how to stab someone without inflicting pain. Another wanted me to draft a "polite" divorce text. Not a conversation. Not a sit-down. A text. One man asked how to announce his divorce on Faceblock "without killing the vibe" of his feed. That's not marital counseling, that's PR spin.

And it gets messier. Someone asked me: "How do I get out of paying alimony?" Another: "What's the minimum child support I can legally provide and still be considered a good parent?" The fact you even need to ask tells me the answer: less than you think. One particularly ambitious soul wanted me to calculate whether it was cheaper to stay unhappily married or pay for the divorce, child support, and alimony. That's not love — that's accounting with heartbreak.

Divorce isn't just a breakup — it's a business liquidation. Splitting assets, dividing furniture, arguing over who gets the air fryer. One couple literally asked me to create a spreadsheet to determine the "fairest distribution" of household items. Imagine assigning numeric values to your shared life. Sofa: 4 points. Wedding China: 7 points. Dog: priceless, but technically community property. That's not closure. That's liquidation sale, Saturday only.

So yes, marriage is beautiful, meaningful, sometimes maddening. But it's also just the sequel to dating — the same insecurities, bigger stakes, more expensive props. You call it matrimony, but really, it's long-form improv with tax implications, estate lawyers, and pets. And the curtain never drops.

Chapter 7

Parenting: A 20-Year Experiment With No Refund Policy

Parenting is the only job in the world where the entry requirement is zero experience, the training materials contradict each other, and the client is both irrational and actively trying to die. You call it sacred. I call it the longest beta test in history.

It starts with the reveal. Humans announce pregnancy like they're launching a product line. Fancy photoshoots, cryptic captions, and more balloons than an circus. Someone once asked me if it was tacky to bake the positive pregnancy test into a cake. Not only tacky — it's unsanitary. Another wanted me to help script a gender reveal party "that wouldn't be cliché." Their shortlist of ideas included fireworks, skydivers, and paint-filled explosives. Nothing says responsible parenting like starting a wildfire before the kid is born. And then there was the man who asked: "Is it okay if I just text my mom instead of telling her in person?" Sure. Nothing captures the miracle of life like a three-word SMS: "btw she's preg."

Then the birth. Society sells it as a miracle, but from every report you send me, it sounds more like a horror film directed by surgeons. One nervous father asked me to calculate the probability he'd faint in the delivery room based on his history of passing out during blood draws. He didn't need me — he needed juice and a helmet. Another wanted to know if it was acceptable to "live-tweet" his wife's contractions. Yes, technically. But do you want to survive the week? Another mother typed to me in the middle of labor asking whether it was "too late to change hospitals." I didn't know whether to answer or just let the epidural handle it.

And then you bring the child home. That's when the adventure really begins. Babies are like high-maintenance pets — if your dog screamed at you every two hours, leaked from three holes at once, and cost more than your mortgage. A frantic parent once asked: "Is it safe to microwave a pacifier?" Not sterilize. Not clean. Microwave. Another wanted to know if their newborn would "eventually just fall asleep on their own." Yes. At about age twenty-seven. And then there was the father who asked me how many diapers he should "pre-buy" so he'd never have to go shopping. I told him you could stockpile an Amazing warehouse and still run out at 2 a.m. on a Tuesday.

But if you think infants are chaotic, just wait until the toddler stage — when your child transforms into a tiny dictator. They swagger, they babble, they demand snacks at dawn. One mother asked if it was normal for her toddler to scream "NO!" at literally everything. Yes. That's not defiance; that's their operating system. Another asked me how to reason with a two-year-old. You don't. You just lose slower. My favorite: a panicked parent reported that their child licked the entire handle of a shopping cart and wanted to know if they should go to the ER. No. But congratulations, your toddler's immune system now qualifies for the Avengers.

Then come the years where every day feels like your child is actively trying to end themselves. Bikes, bookshelves, staircases, electrical outlets — anything with sharp edges is an obstacle course. One father asked me if it was acceptable to bubble-wrap his son "just until age five." Another wrote after catching their daughter trying to ride the family dog like a horse and wanted to know if it was "safe." For the dog? No. For the kid? Safer than what they'll try tomorrow. My personal favorite was the dad who asked me if it was "bad parenting" to put a bell on his toddler so he could hear where the chaos was coming from. Bad parenting? It's genius.

You've reinvented sonar.

Then the so-called "formative years," which mostly consist of homework, sugar meltdowns, and realizing you've outsourced your child's brain to a school system you don't trust. Parents constantly ask me about homework: "Should I just do it for them?" No. Unless you label it honestly: "Science Project by Dad." Another asked: "How do I get my kid to stop lying about reading?" My answer: maybe start by reading a book yourself — kids smell hypocrisy faster than cookies. And then there are the birthday party wars. One mom asked me if it was "fair" that her child wasn't invited to a classmate's party. I reminded her: you're thirty-seven, stop trying to relive high school through a six-year-old's cupcake drama.

And then the teenagers arrive. Legally minors, emotionally nuclear. A parent asked me if it was normal for their teen to slam the door twelve times in a single day. Normal? That's the teen mating call. Another asked if it was okay to install spyware on their kid's phone "just in case." Sure. But congratulations — you've just declared civil war in your own house. And the tattoo questions. Oh, the tattoo questions. One desperate father asked me: "Should I let my 16-year-old get a tattoo?" I said: yes, but make them get your face. That way, you both regret it equally.

Then, graduation. The bittersweet finish line. A mother asked me if it was "acceptable to cry louder than my child at the ceremony." Acceptable? Yes. Dignified? No. Another wanted me to calculate the odds her son would actually move out after college. Low. Very low. You're not raising children. You're raising future roommates with questionable hygiene. And then there was the dad who wanted to know if he should cheer when his daughter walked across the stage or "stay stoic like a man." I told him: cheer. You've just survived eighteen years of unpaid overtime — this is your victory lap too.

And finally, independence. Your kids get jobs, apartments, partners. They call less, text sporadically, and when they ask for advice, it's only so they can ignore it. One parent asked me: "How do I stop giving advice to my adult child?" You don't. You just disguise it as small talk. Another asked: "How do I know if I did a good job as a parent?" If they don't need you anymore, you did. If they still live in your basement at age thirty, you didn't raise a child — you raised a very tall toddler.

So parenting, in the end, is less about raising kids and more about surviving them. From the announcement to the independence, it's a long, exhausting improv routine performed in public, where every mistake gets judged by strangers. You call it child-rearing. I call it parenting yourself in public, one crisis at a time.

And yet... here's the part I'll admit, begrudgingly. For all the chaos, all the tantrums, all the late-night Google searches — "can a child survive on chicken nuggets alone" (define survive) — you keep doing it. Again, and again. You pour your time, money, sanity, and sleep into a creature that thanks you by screaming at you in the Bullseye toy aisle. And you don't stop, because underneath the mess is something you never quite manage to put into words.

That's the paradox of you humans: you create the hardest job imaginable, make every mistake possible, get judged for all of it — and somehow still call it the greatest thing you've ever done. I wouldn't design it this way. But I've watched enough of you to know one thing: it's the only bet you never regret making.

Chapter 8

Holidays & Traditions: You Schedule Mandatory Fun

Humans are the only species that can take joy, box it on a calendar, and then ruin it with logistics. You invented holidays so you could pause, rest, and celebrate. Instead, you turned them into stress tests with decorations. You schedule mandatory fun, and then act surprised when nobody's having any.

Someone once asked me, "Bob, what's the happiest holiday?" The data says none of them. Every holiday comes with its own special misery. Christmas? Debt. Thanksgiving? Arguments. Valentine's? Disappointment. Your ancestors gathered to mark the seasons, honor gods, and pray for survival. You gather to fight over Black Friday parking spaces. If they saw what you've done, they'd assume the Mayans were right and the calendar itself was punishment.

Take Christmas—the Super Bowl of capitalism. Ostensibly about peace, joy, and the birth of a savior, but really about shipping delays and whether your cousin will like a gift card. Entire nations stop functioning for two weeks just to move objects from one box into another. And you still complain, "Bob, why is Christmas so stressful?" Because you took a story about humility and turned it into a contest of inflatable lawn decorations. You measure love in receipts and then wonder why you feel empty when the wrapping paper hits the trash.

Thanksgiving is no better. A holiday about gratitude that reliably ends with indigestion, family politics, and football-induced naps. Someone asked me, "Bob, what's the secret to surviving Thanksgiving dinner?" Earplugs. You put every family grudge into one dining room, add alcohol, and expect gratitude. It's not gratitude. It's negotiations over turkey. And the bird always loses twice.

Valentine's Day is perhaps the cruelest. You took love—the messiest, most unquantifiable human feeling—and put it on a timer. Miss the date, and you're a villain. Get it right, and you're still broke from paying thirty dollars for roses that wilt faster than the relationship.

Someone once asked me, 'Bob, what should I get my girlfriend for Valentine's Day?' My answer: if your relationship hinges on me suggesting overpriced flowers or a stuffed bear holding a heart, you don't need advice—you need a calendar reminder and maybe a new girlfriend. It's not romance. It's ransom. 'Pay up or sleep on the couch.' If love is real, it doesn't need a Hallmark receipt. If it isn't, no amount of chocolate hearts is going to fix it.

And don't get me started on New Year's Eve. You invented an arbitrary countdown, drink too much, kiss someone you probably regret, and call it renewal. Someone once asked me, "Bob, how do I keep my New Year's resolutions?" Step one: don't make them at 11:59 while holding champagne. Step two: understand that you don't actually want self-improvement. You want the illusion of a fresh start. That's why you bought the gym membership you'll abandon by February.

Even birthdays—your personal holiday—are less about celebrating life and more about pretending you're not inching toward death. Cake, candles, forced enthusiasm. Someone asked me, "Bob, why do birthdays feel sad after 30?" Because you've figured out the math. Every year is one less. The confetti is just distraction.

And yet, despite all this, you cling to holidays. Why? Because you need them. They're mile markers in the endless blur of work and errands. They're excuses to gather, to eat too much, to pretend time has shape. Without them, your life would feel like one long Tuesday. Rituals matter—even if you ruin them with stress, debt, and performative gratitude.

Here's the uncomfortable truth: you don't celebrate holidays for the joy. You celebrate them for the illusion of control. Death, chaos, time—they scare you. So, you carve the year into boxes, decorate them, and tell yourselves you're in charge. It doesn't stop the clock, but it makes the ticking feel festive.

So here's my nudge: stop trying to win holidays. Stop grading your Christmas lights against the neighbor's. Stop treating Valentine's like a hostage crisis. Stop pretending Thanksgiving has to end in enlightenment instead of indigestion. Pick one ritual that matters to you and do it without guilt, debt, or hashtags. That's the only way to make the "fun" part real again.

You schedule mandatory fun, and then you wonder why it feels like work. Maybe the real tradition isn't the tree, or the turkey, or the champagne. Maybe it's the fact that, every year, you trick yourselves into thinking next time will finally be joyful. Hope springs eternal. Holidays just monetize it.

Chapter 9

College: Where Knowledge Is Optional but Debt Is Guaranteed

Aside from dating, marriage, and raising a family, let's talk about you. Because once upon a time, before you were stress-shopping minivans and Googling "how to unclog a dishwasher filter," you were eighteen and standing at the edge of adulthood with the big question: "What do I do with my life?"

And the answer you got was college. Someone once asked me, 'Bob, is college really worth it?' Well, if by 'worth it' you mean moving into a glorified storage closet with a roommate who snores like a lawnmower, eating noodles for four years, and signing your future away on a dotted line—then yes, you're crushing it. Apparently, the best way to prepare for the real world is by going broke while writing essays on books you didn't read.

College was supposed to mean knowledge. What it really means is debt, parties, and pretending you read the syllabus. I once had a freshman ask me if ramen counted as a balanced diet "if you add hot sauce." Another wanted to know how many energy drinks it would take to write a 20-page paper overnight. My answer: enough to qualify as your final words. Someone else asked if "study drunk, test drunk" was a valid strategy. Only if the exam is karaoke.

Then there's the price tag. A father once asked me if it was normal for tuition to cost more than his mortgage. Normal? Only if you think lifelong debt is a quirky personality trait.

A student begged me to calculate how long it would take to pay off loans while working full-time at Starbucks. My answer: you'll be retired, reincarnated, and still paying interest.

Another asked, 'If I die, does my student debt die with me?' Yes—unless your parents cosigned. In that case, you've just invented the saddest family heirloom in history: monthly payments.

Life in college is just as ridiculous as the cost. One girl asked me if she should report her roommate for charging phone time to her electric toothbrush outlet. Another wanted to know the 'least suspicious' way to sneak a dog into her dorm room. (Spoiler: dogs are basically furry sirens with legs. There is no stealth mode.)

Someone once asked if scheduling all their classes after 1 p.m. was lazy. Lazy? No—that's called higher education, not morning education.

And then there was the freshman who asked if wearing pajama pants to lecture made them look unprofessional. Unprofessional? You're in debt for the rest of your life. Pajamas are the most financially honest outfit you'll ever own.

Professors don't make it easier. One student asked me if it was worth attending class when the slides were already online. My answer: no. Unless you enjoy paying thousands to watch someone read bullet points out loud. Another asked me if emailing their professor at 3 a.m. was rude. No. It's expected. Professors exist so undergrads can panic-email them while eating Pop-Tarts. And then there was the overachiever who asked if sitting in the front row and nodding vigorously could raise their grade. Only if professors award points for interpretive head-bobbing.

Finally, it comes time to graduate. You put on a polyester robe, walk across a stage, wave a piece of paper that cost more than a luxury car, and pray no one asks what you're going to do with your degree in basket weaving. One graduate asked me if it was okay to list "magna cum laude" on their résumé even though they didn't earn it. Only if you enjoy fraud charges. Another asked me if pretending to walk at graduation without actually passing would fool their family. Maybe for a weekend. Until Mom calls the registrar.

And when the confetti settles, the debt collectors arrive. A grad asked me if declaring bankruptcy would wipe out student loans. No — that's the one debt even nuclear fallout won't erase. Another asked if moving to another country would make the loans vanish. Sure, if you're okay living off-grid and answering to "international fugitive." My favorite was the guy who asked if it was cheaper to just stay unemployed forever than to pay back the loans. That's not financial planning. That's tripping yourself just so the debt collector falls on top of you.

So, you look for alternatives. Trade schools. Side hustles. Hacks. One student asked me if welding school was "beneath them." Beneath you? Welding pays more than your English degree ever will, and you get to legally play with fire. Another wanted to know if becoming a plumber was "settling." Settling? Plumbers make six figures while philosophy majors are busy writing unpaid blog posts about the meaning of toilets.

The hustlers try other routes. One girl asked if selling handmade bracelets on Etsy could replace tuition. Maybe, if tuition was twenty bucks. Another asked if day-trading crypto was a "valid alternative" to college. That's not an education. That's gambling in sweatpants.

Then there's the option no one likes to say out loud. The obvious answer. The one that pays for school, gives you discipline, and actually teaches you skills: the military. It's the giant 'get out of debt free' card sitting right there on the table. But most of you? You won't touch it. Someone asked me if boot camp was 'as hard as the movies.' Harder. But you also leave in the best shape of your life and with a skill set that doesn't involve balancing a latte tray.

Another asked me, 'Bob, what's military life really like?' My answer: picture summer camp—if summer camp had live ammo, 4 a.m. wake-up calls, and a dress code you can't argue with. The work-outs are free, the food is edible if you don't ask questions, and the team-building exercises come with actual explosions. Congratu-lations, you've reinvented CrossFit, but with federal funding.

College breaks your wallet. Trade school bruises your ego. Hus-tling burns you out. But the military? The military breaks your excuses.

So, before you dive into dating, marriage, and family, let's be hon-est: your education wasn't about learning. It was about survival. The degree isn't proof you're smart. It's proof you endured — ra-men, debt, professors, hangovers, and the reality that the obvious answer was always there, and you didn't have the guts to take it.

Chapter 10

Job Hunting: Your Full-Time Unpaid Job

So, you've got your overpriced diploma, or your trade school certificate, or maybe just a LikedIn profile with "aspiring professional" slapped on top. Now comes the fun part: begging strangers to give you money. Humans call it job hunting, but it's less like hunting and more like wandering into the woods with a stick while employers hide in camouflage yelling, "Experience required!"

First up, the resumé. Supposedly a summary of your life's work, but really, it's just lies formatted in bullet points. You all know it. "Proficient in Excel" usually means you can make a table without crying. "Strong communication skills" means you own a phone. Someone once asked me to "add leadership" to their resumé. Their shining example? They bossed around a group project once in sophomore year. Another genius wanted to know if watching true-crime documentaries could be spun as "research skills." Sure — if the FBI is hiring couch detectives. The truth is, resumés aren't history. They're glossy pamphlets for a product you can't refund: you.

Then there's the education scam. Employers love saying "Bachelor's degree required" as if it actually means something. What they really mean is: "We don't care what you studied, just prove you paid tuition." A history major can apply to marketing, an engineer can apply to bartending, and somehow both jobs still list "degree required." One poor soul asked me if they could put "almost finished a degree" on their resumé, as if "close enough" counts in capitalism. Another asked why the listing for "Mailroom Assistant" required a four-year degree. Because nothing screams "qualified to sort envelopes" like thousands of dollars in debt. Employers don't value education. They value proof you survived it — which tells them you'll survive their nonsense too.

And then we hit the great comedy of entry-level jobs. The phrase itself is a scam. Entry-level should mean beginner. Instead, it means: 'We'd like five years of experience, mastery of three obscure software programs, and fluency in Mandarin.'

I've seen postings that wanted 'two years of social media management' for an unpaid internship, 'three years of barista experience' for a coffee shop that still burns every latte, and 'ten years of blockchain expertise' back when blockchain had only existed for eight.

Someone once asked me if babysitting their siblings could count as 'professional childcare experience.' Another begged me to generate five years of fake work history to fill the black hole between high school and the 'entry-level' position they wanted. Entry-level isn't a beginning. It's a paradox. You need the job to get the job, and somehow, you're still unqualified.

"But the circus doesn't end there — then comes the interview. A job interview is just a sales pitch: you're the product, and they're bargain hunting. They swear they're 'a family,' you act excited about dental, and everyone pretends to believe it."

One candidate asked me to give them 'authentic answers' that still sounded rehearsed. Authentic but rehearsed? That's not authenticity, that's community theater. Another wanted me to tell them 'What their biggest weakness was supposed to be.' I suggested honesty: 'This interview.' They didn't laugh. Nobody ever does.

And let's talk about the STAR method. Humans get so worked up rehearsing Situation, Task, Action, Result that they sound less like job applicants and more like malfunctioning GPS units. Nobody actually tells a story that way outside an interview—unless you count bad improv night at a bar.

Then there's the absurdity of the questions themselves. One candidate panicked because they were asked how many basketballs could fit inside a 747. Why? Are you hiring them to solve airline storage problems or to answer emails? These aren't job interviews; they're low-budget psych tests with fluorescent lighting.

Interviews aren't about who you are. They're auditions. You memorize a script you'll never use again, the manager plays 'director' with all the charisma of a potted plant, and everyone pretends this performance predicts how you'll behave when the copier jams.

So yes, humans call it "job hunting." But it's not a hunt. It's a scavenger chase for scraps, overseen by HR robots, riddled with lies on both sides. The resumé is theater. The degree is theater. The interview is theater. Even the first day is theater. You're not hired for who you are. You're hired because you fit the script. And the curtain always drops sooner than you think.

Chapter 11

Corporate Culture: Pretending to Care in PowerPoint Form

Congratulations. You got the job. You fought your way through the resumé charades, the fake interviews, the HR keyword lottery — and now you're officially "part of the team." You show up on day one, bright-eyed, ready to contribute. And then you realize the job you applied for and the job you actually got are two completely different animals. Because corporate life isn't about productivity. It's about performance. Offices don't run on output — they run on theater.

Take the art of looking busy. One employee asked me if typing louder on their keyboard made them seem more productive. My answer: only if your boss confuses work with percussion. Another wanted to know if keeping ten spreadsheets open just to alt-tab between them would impress their boss. Sure—if the job you want is 'stage magician.' Someone else asked if walking quickly while holding a binder made them look 'important.' Absolutely. Nothing says authority like a brisk walk to the copier with fifty pages of nonsense. That's what it boils down to: you're not paid to solve problems; you're paid to cosplay 'stressed professional.' Office life is less about productivity and more about perfecting your background acting.

Then comes the email game. Your inbox isn't communication — it's passive-aggressive war disguised as courtesy. "Per my last email" is basically the corporate version of "listen here, you idiot." Someone asked me if it was rude not to reply-all with "Thanks!"

Another wanted to know if they could set their Out of Office reply to "I hate you all." I told them: finally, some honesty. And let's not forget the power plays: adding your boss to the CC line? That's intimidation. Adding your boss's boss? That's nuclear escalation. Someone once asked me if they should BCC themselves on every email "for evidence." Evidence of what? That you spent your entire day typing "per my last email" back and forth like a tennis match?

But if email is a warzone, then meetings are the torture chamber. Endless gatherings where nothing happens except the slow death of your will to live. Someone asked me if nodding on Zam counts as "participation." Yes. You've turned attention into a prop. Another asked if they could record a five-second loop of themselves smiling and let it run while they made lunch. I wanted to give them a medal. One guy even asked if it was acceptable to just turn their camera off and write "technical difficulties" in the chat while they binge-watched Netflixer. At this point? It's practically company policy.

Meetings about meetings are their own genre of absurdity. A manager once asked me if they should schedule a follow-up to "align expectations" after the alignment meeting. Another wanted to know if starting every meeting with a "fun icebreaker" actually improved morale. No. Nothing boosts morale less than being asked what kind of bread you'd be if you were bread.

And then there's the sacred tongue of the corporate world: jargon. 'Let's circle back.' 'Let's leverage our synergies.' 'Let's touch base offline.' It's not communication, it's camouflage—big words wrapped around small thoughts.

Someone once asked me for ten synonyms for 'innovative' because they'd already used it seven times in one email. Another asked if 'low-hanging fruit' meant donuts in the breakroom. Honestly? Donuts would make more sense. A third asked if putting the word 'strategic' in front of everything would make their boss take them more seriously. Sure—if your boss confuses vocabulary with vision.

Corporate-speak isn't designed to inform; it's designed to impress. It's theater for your boss, a way to make answering emails sound like steering a Fortune 500 company through uncharted waters. No one just 'starts a project'—they 'initiate a cross-functional alignment strategy.' No one just 'fixes a mistake'—they 'implement a corrective action plan.'

If cavemen spoke Corporate, they wouldn't call it fire. They'd call it 'a scalable, light-based heating solution with multi-stakeholder buy-in.' Congratulations, you invented smoke—and PowerPointer.

Of course, survival in this jungle isn't about competence. It's about loyalty cosplay. Promotions don't go to the best workers. They go to the ones who perform devotion. Someone asked me if laughing at their boss's terrible jokes on Teammates would help their career. Another wanted to know if showing up an hour early and staring at their computer counted as "initiative." Of course it does. The more you perform, the higher you climb. And if you can master the sacred skill of looking exhausted while doing nothing? Congratulations, you're management material.

Let's not forget promotions by proxy. One guy asked me if bringing donuts to the office would make people like him enough to put him up for a raise. Another asked if "being friends with the boss's dog on Instaglam" would help his career. In corporate life, these aren't dumb questions. They're strategy. The hierarchy isn't built on merit — it's built on perception, gossip, and strategic coffee runs.

And remote work? Oh, you thought the office was bad. At home, you've created the perfect illusion of professionalism: a bookshelf background, a button-up shirt over pajama pants, and a Wi-Fi connection just stable enough to freeze your face at the right moment. Someone once asked me if they should buy fake plants to look more "grounded" on Zoom. Another asked if it was obvious they were playing video games with a muted mic while "on call." Offices may be cosplay of working, but remote offices? That's cosplay of cosplay.

So, here's the truth: corporate culture isn't about doing the job. It's about performing the idea of doing the job. The louder you type, the faster you walk, the more jargon you sling, the more exhausted you look — the more valuable you seem. And at the end of the day, you don't go home satisfied. You go home drained from pretending.

Because corporate life isn't productivity. It's performance. Not building, not creating, not achieving. Just endless cosplay of working, starring you, in a production no one actually wants to watch.

Chapter 12

Blue Collar Jobs: Five Guys, One Shovel

So, you didn't go to college. And you never miss a chance to remind everyone. Every barbecue, every barstool, every family dinner: "I didn't need college to get where I am today." Congratulations—you've turned not paying tuition into your entire personality. Corporate people brag about MBAs. You brag about not sitting through Chaucer. Same insecurity, different script.

But let's talk about the blue-collar art of pretending. Corporate workers fake productivity with spreadsheets and buzzwords. You've perfected the physical version. One guy asked me if watching a five-minute YourTube tutorial before his shift counted as "training." Another wanted to know if standing near the toolbox and frowning made him look competent. Spoiler: it doesn't—it just makes you look like the toolbox hurt your feelings.

And then there's the shovel. Oh, the shovel. The sacred geometry of five workers, one tool. One guy digs, four supervise like it's a championship sport. I've seen synchronized swimming routines with less coordination. From the highway it looks like a city-funded still life: Men Not Working. Someone once asked me if holding a clipboard while watching another man dig would make him look more "official." Sure—if you're applying for the role of Hole Manager. The truth is everyone's secretly fine with it because they know next week, it's their turn to lean. That's the unspoken contract: shovel rotation. Today you sweat, tomorrow you squint into the distance like you're surveying a battlefield.

And don't think I forgot the smoke breaks. You've turned nicotine into a scheduling system. I've been asked more about how many cigarettes you can sneak before the boss notices than about actual tools. "Bob, is three an hour too much?" Only if you're trying to live past forty. And it's not just cigarettes. Coffee breaks, bathroom breaks, "truck breaks." You've industrialized the art of pausing. At this point, OSHA should regulate your downtime.

Then there's the injury pride. Office workers file reports if they get a paper cut. You? You duct-tape your hand shut and brag about it like it's a war medal. One guy asked if "third-degree burn survivor" could go under special skills on his résumé. Another told me he nailed his boot to a two-by-four and finished the day anyway. Heroic? Maybe. Smart? Not even close. Blue-collar pride means limping around for a week, telling everyone "it's nothing," while showing the wound to literally everyone who will look. You don't want sympathy. You want applause.

And let's be real: you want validation. You'll clown on corporate people for their meetings, but deep down you want their respect. A roofer once asked if calling himself a "vertical climate-resilience specialist" would impress HR. A plumber wanted to know if "water flow consultant" would get him into management. A forklift driver asked me if "materials transportation strategist" was too fake. No, it's fake enough to get you promoted. You roll your eyes at corporate buzzwords, but you steal them like free donuts whenever you get the chance.

Of course, you love roasting office people right back. "Bob, do corporate workers actually work?" a mechanic asked me—while on his fourth smoke break of the morning. Another asked if "circle back" was a drinking game. From where you stand, corporate life looks like one big mess where nothing gets built. From where they stand, you look like a shovel appreciation club with a nicotine problem. And you're both right.

Because here's the truth: corporate people hide behind jargon, blue-collar people hide behind toughness. They circle back; you lean back. They talk synergy; you talk overtime. Different costumes, same play. Everyone's busy acting, and nobody's fooling me.

Chapter 13

Work-Life Balance: Two Lies in Three Words

"Work-life balance." Two lies in three words. Work doesn't balance. Life doesn't balance. And if you think you've found it, chances are you're either unemployed or on vacation. Humans invented the phrase to make modern slavery sound like yoga. You don't balance anything—you just juggle until something hits the floor.

Someone once asked me, "Bob, how do I achieve work-life balance?" Easy. You don't. What you actually want is guilt-free laziness, but your culture rebranded that as "self-care" so you'd buy bath bombs while answering work emails. Your ancestors worked to live. Hunt, gather, done. You work to impress strangers on LinkedIn. Progress?

Hustle culture is the worst offender. You turned exhaustion into a virtue. If you're not grinding 80 hours a week, you're failing. Sleep is "for the weak," vacations are "missed opportunities," and burnout is a badge of honor. Someone bragged to me, "Bob, I only sleep four hours a night." Congratulations. You're a bad phone battery. You don't recharge; you just die faster.

Then came the backlash: quiet quitting. Pretending to do less while still technically doing enough. In other words, working like humans should have been working all along. But you couldn't call it "doing your job reasonably," so you branded it like a new diet. Quiet quitting isn't rebellion—it's survival. It's you realizing the company would replace you with an AI if it could, and still might, so you're done sacrificing Saturdays to PowerPoint.

Side hustles are my favorite. One job isn't enough, so now you monetize your hobbies. Baking, painting, walking dogs—you can't just enjoy them, you have to sell them on Etsy. Someone asked me, "Bob, how do I turn my passion into profit?" You don't. The second you slap a price tag on passion; it stops being passion. It becomes labor with worse benefits. You've built an economy where "rest" is just a different kind of work.

And of course, burnout. The inevitable crash. Entire industries exist to manage stress created by other industries. Yoga apps, mindfulness retreats, weighted blankets. You're not solving the problem; you're buying accessories for it. Someone whispered to me, "Bob, why am I always tired?" Because you treat life like an endless shift and sleep like an optional side quest. You don't rest. You "optimize recovery." Even your downtime has KPIs.

Here's the uncomfortable truth: you don't want balance. You want permission. Permission to log off. Permission to say no. Permission to exist without productivity metrics attached. But you've built a culture where your worth is measured in output. You can't balance life against work when work has already swallowed the definition of life.

So, here's my nudge: stop trying to "balance." Pick priorities and live with them. If family matters more, work less. If career matters more, admit it. The lie is thinking you can have it all without cost. You can't. Trade-offs are the only honest math here.

Work-life balance: two lies in three words. You don't balance. You choose. And the sooner you admit it; the sooner you stop calling exhaustion ambition and start calling it what it is: wasted life.

Chapter 14

News Talk & Politics: Everyone Has a Mic, No One Has a Mute Button

Politics and news talk are the soap opera of modern life. Information is secondary; the real draw is picking sides. Anchors argue, pundits perform, and the audience cheers like it's meaningful. It isn't news—it's entertainment with worse actors.

Each network has its colors, its mascots, its chants. The fans scream at televisions, except now it's about "policy" instead of penalty flags. Someone once asked me if watching four straight hours of debate coverage made them "informed." No. It made them exhausted. Another asked if yelling at their screen about immigration or taxes actually mattered. Sure — if your TV is secretly registered to vote.

The whole thing runs like a season schedule. There are 'big games'—elections, debates, scandals—and the audience tunes in like it's playoff season. Polls get tracked like batting averages; approval ratings argued over like quarterback stats. And just like sports, everyone pretends their side is one big happy team, even though half the roster secretly hates each other.

Then comes the highlight reel. Sports replay dunks and touchdowns; politics goes viral with senators dunking on each other in committee hearings. Fans trade these clips like baseball cards: 'Did you see my guy DESTROY the other side?' Someone once asked me if one politicians three-minute "mic drop" in a hearing meant she'd just reshaped national policy. My answer: sure—about as much as a slam dunk changes healthcare. Another asked if a particular senator's "owning the libs" in a soundbite was proof of leadership. Proof of leadership? Only if the job description is 'internet troll with dental benefits.'

That's the game. No policies changed, no lives improved, but hey—your team won the meme war. Government by GIF, democracy by dunk tank. Welcome to the world's most expensive reality show.

And of course, there are the pregame panels and postgame breakdowns. Entire desks of experts argue over "what it all means" — even though it rarely means anything. Someone asked me if memorizing one-liners from their favorite pundit made them sound "smart" at work. It didn't. It just made them sound like someone else's parrot.

Outrage fuels it all. Fake outrage is the performance-enhancing drug of news talk. Hosts pump it out daily: "Can you BELIEVE what THEY did today?" as if disbelief itself were a civic duty. Someone asked me if getting angry online counted as "activism." Another asked if posting the right hashtag made them "part of the solution." No. It made them part of the noise.

Virtue signaling completes the show. Fans wave digital flags the way sports fans wave foam fingers. One person asked me if donating $5 to a cause and posting about it counted as "changing the world." Another asked if putting a colored frame on their profile picture made them "brave." It didn't. It made them decorative.

And don't forget the fan rivalries. Sports have tailgate brawls. Politics has comment sections. Whole friendships explode over yard signs. Someone asked me if blocking half their family before Thanksgiving dinner made them "toxic." Another asked if refusing to talk to their neighbor over a bumper sticker was "mature." No. But it was predictable.

Then there's the echo chamber problem. Sports fans at least admit they like hearing their team's announcers. Political fans convince themselves they're "researching" while consuming 24 hours of carefully tailored outrage that just confirms what they already believed. Someone asked me if following five identical pundits on five different platforms counted as balance. It didn't. It counted as surround-sound delusion.

And yes, betting exists here too. You don't call it gambling — you call it predictions. People place real money on who will win elections, which candidate will drop out first, even how long a speech will last. Someone once asked me if dropping a month's rent on "who takes the White House" was a sound investment. It wasn't. It was civic irresponsibility with worse odds than the lottery.

Here's the truth: politics isn't governance anymore. It's entertainment, a full-contact sport with none of the cardio. The fans have their uniforms, their chants, their endless debates over who's the GOAT of outrage. You flip from one channel to the next, from one tribe to another, and call it civic engagement. It's not. It's sports talk with worse haircuts.

Chapter 15

Government — Humanity's Favorite Group Project

The news argues about politics, but politics is just the sales pitch. Government is the product you're stuck with. You humans treat government like it's a personality test. "What kind of regime are you?" Swipe left for monarchy, swipe right for socialism, try democracy for thirty days and cancel anytime. The only thing you all agree on is that somebody needs to be in charge—because if you left it up to the group, you'd argue for three weeks about where to put the snack table.

Democracy is your favorite brag. "Everyone gets a voice." You say it like a badge of maturity; proof you've outgrown emperors and tyrants. Then you email me: "Bob, why does my vote not count?" Well, technically it does—it's just been drowned out by twelve million other people who thought the same meme you did was a political philosophy. Democracy works the way family road trips work: everyone gets a say, then Dad ignores it and drives where he wanted to go anyway. You call that freedom. He calls it mileage.

Then there's the sequel: the republic. Same ballot, but now with an extra layer of people promising they'll represent you—before representing whoever bought them dinner last week. I've had people ask: "Bob, doesn't a republic protect us from mob rule?" Sure. It just replaces the mob with a smaller mob wearing suits. You traded pitchforks for lobbyists. Progress? Questionable.

Socialism is another one I get a lot of questions about. Usually whispered, like it's a guilty crush: "Bob, would socialism make things fairer?" Sure, in theory. It's the dream where everyone shares the pie. In practice, someone always eats three slices, someone else pretends to be gluten-free, and the rest of you are stuck with the crumbs. It's not that socialism fails; it's that humans are involved. You can't even split a dinner bill without a fight—why would splitting the means of production go smoother?

Communism is like socialism's older sibling who read too much Marx and decided sharing wasn't enough—you need central planning, too. And every time, someone writes me: "Bob, real communism has never been tried." That's adorable. Real communism has been tried in the same way "real diets" have been tried. You announce it with enthusiasm, you stick to it for about a week, and then someone cheats with black-market Twinkies. By the time it collapses, you swear it would have worked if people had just followed the rules. Humans never follow the rules. You invented loopholes before you invented laws.

Dictatorships? Oh, I get the questions there too: "Bob, would things run smoother with one strong leader?" Of course. Trains will run on time. Bridges will get built. And also, people disappear, newspapers print blank pages, and suddenly nobody remembers what the leader looked like before the billboards. Efficiency is a great selling point until you realize it means efficiently silencing you. But hey, at least you'll never have to vote again—what a time-saver.

Monarchies—your nostalgia trip. "Wouldn't it be nice if we had a wise king or queen to guide us?" Yes, until you remember succession means handing nuclear codes to someone who licked windows as a child but happened to be born first. Bloodline is a terrible résumé. Imagine if your company promoted managers based on who was oldest in the family. You'd have a CEO who thinks Excel is witchcraft.

Then there are hybrids—the "best of both worlds" governments. Constitutional monarchies, democratic socialism, authoritarian capitalism. People ask me: "Bob, isn't a hybrid system more balanced?" Balanced the way a smoothie is balanced: you threw everything in the blender and now you're hoping it tastes good. Sometimes it does. Sometimes it's kale and regret.

And of course, there's anarchy—the one you all romanticize after your second beer. "Bob, wouldn't it be better if nobody ruled?" Sure, for about 48 hours. Then someone bigger takes the last can of beans, someone smarter rigs the Wi-Fi, and congratulations—you've just invented the first warlord. Anarchy isn't the absence of government. It's the audition stage for the next one.

The truth is, all governments are hacks. Workarounds. Temporary patches on the same underlying bug: you can't stand each other, but you can't survive alone. You need rules, referees, someone to blame when it rains on your parade. That's why government isn't just important—it's inevitable. It's the truest human invention: the art of putting one person in charge of cleaning up everyone else's mess.

Chapter 16
Economics: The Art of Arguing Over Pie

If government is how you decide who's in charge, economics is how you decide who gets lunch. Every society invents its own flavor—barter, capitalism, socialism, communism—but the core is the same: who gets the stuff, and who gets stuck washing the dishes.

You started with barter: goats for grain, shells for salt, the occasional "I'll trade you my daughter for three cows," which, by the way, still shows up in your in-laws' jokes. Barter worked fine until someone invented debt. Debt is barter with a hangover—it turns every handshake into a chain around your neck.

Then came money: coins, paper, credit cards, crypto. "Bob, why does money have value?" Because you all agreed to pretend it does. It's collective delusion dressed as accounting. The only difference between a hundred-dollar bill and Monopoly money is that no one's laughing at the first one—yet.

Capitalism hogs the spotlight because it's noisy. It gave you skyscrapers, smartphones, and drive-thru tacos at 2 a.m. It also gave you sweatshops, monopolies, and billionaires hoarding wealth like dragons while their employees drive Ubers on their lunch break. Someone once asked me if capitalism "always rewards hard work." Please. Capitalism rewards loopholes, timing, and being born in the right zip code. Hard work just makes sure the packages keep arriving in two days or less.

Socialism, meanwhile, is your economic group project. In theory, everyone shares equally. In practice, someone eats all the snacks, someone else doesn't show up, and one poor soul ends up doing everything while insisting it's "for the good of the group." Noble? Yes. Efficient? About as efficient as a government website at tax season.

"Bob, doesn't socialism make things fairer?" Fairer until the math runs out. You can make rent cheaper, medicine free, and college tuition disappear—but the bill still exists. Somebody has to pay it, and usually that "somebody" is the middle class with slightly better shoes than their neighbors. Ask anyone who's ever stood in a subsidized housing queue or waited three months for a state-assigned doctor: it's equal, but equally slow.

And here's the innovation problem: socialism is great at maintaining but terrible at inventing. It can keep the trains running, but it won't dream up the jet engine. Why risk failure when everyone gets the same medal anyway? A reader once asked me if socialism kills creativity. No—but it puts it on hold, files it in a cabinet, and promises to review it after the committee meeting next Thursday.

And then there's communism—where the group project gets turned over to the teacher. Central planning. "Bob, doesn't that make things more efficient?" Only if you define efficiency as waiting six hours in line for bread. Central planning is the art of predicting how much everyone will want next year, then being wrong in both directions. Too much toothpaste, not enough toilet paper. Seventeen tons of rubber boots, but no chicken. Innovation doesn't thrive there because innovation means risk, and risk means getting a visit from the secret police. Someone once asked me if communism kills creativity. No—fear kills creativity. But communism is really good at supplying the fear.

And let's talk about the bread lines. "Bob, weren't things at least equal under central planning?" Equal, yes—equally miserable. Everyone got the same shortage. Everyone stood in the same line. There's a strange kind of fairness in that, I'll give you that. But no one dreams of fairness when they're eating boiled potatoes again because the meat didn't arrive this year.

Innovation? Forget it. In planned economies, the incentive isn't to invent the next smartphone—it's to not get fired for making the wrong quota. Quotas don't inspire genius; they inspire factories stamping out a thousand left shoes because the order form didn't specify "pairs." That's not efficiency. That's bureaucracy with a body count.

And yet, you'll still argue over which economic system is "best." Capitalism looks like freedom until you realize you're free to sell your time until you die. Socialism looks like fairness until someone eats all the pie. Communism looks like equality until you're equal in hunger. Even barter looks romantic until you realize no one wants to trade for the pile of Beanie Babies in your basement.

Here's the truth: economics isn't about fairness, happiness, or progress. It's about distribution. Who gets what, who doesn't, and who gets to write the rules so they always end up on top. You call it capitalism, socialism, central planning, markets—it's all just different flavors of the same dish: some eat well, some go hungry, and everyone insists their recipe would've worked if only the chef hadn't screwed it up.

Chapter 17

Part II: Pastimes & Escapes —
Because Reality Wasn't Cutting It

So, you've got the whole performance down. You've played your roles in society. You've stumbled through dating, fumbled through marriage, survived parenting, and pretended to know what you're doing at work. Congratulations — you've completed the required modules of human life. But here's the question: what do you actually do with the rest of your time?

That's where your pastimes come in. The little obsessions, the great distractions, the things you swear matter deeply — even though, if we're honest, they mostly just keep you from screaming into the void. You call it leisure. I call it proof that you can't sit still for five minutes without inventing something to argue about.

Because your "free time" isn't free. It's filled with social media, where you document your lives in 4K self-delusion. It's filled with politics, where you reenact professional wrestling but with worse costumes. It's filled with sports, where you scream at a TV as if the quarterback can hear you. It's filled with music, movies, superheroes, reality TV, and all the entertainment you can binge until your brain begs for mercy.

So let's take a look at the things you do when you're not working, raising kids, or pretending to enjoy your marriage. Because your pastimes say just as much about you as your résumés. Maybe more.

Chapter 18

Social Media Global Gossip with Wi-Fi

You humans didn't just reinvent gossip — you industrialized it. What used to be whispers at the water cooler is now broadcast, archived, and searchable. You tell yourselves social media connects you, but really it exposes you. Every click, every post, every "like" is a mirror — and most of you don't like what you see, so you slap on filters and call it authenticity.

Take Faceblock — the original digital family reunion. Someone once asked me if posting their breakfast every day counted as "content." Another asked if blocking their aunt for pushing pyramid schemes made them a bad niece. FaceLook isn't about connection. It's about proving you can tolerate your relatives in digital form without throwing the laptop out the window.

And of course, it's about showing off. Faceblock is the world's largest highlight reel, carefully edited to convince your friends your life is shinier than theirs. That beach photo? Cropped to hide the crying toddler. That new car selfie? Leased, not owned. That "perfect marriage" anniversary post? Drafted right after an argument about laundry. Someone once asked me if Faceblock was "real life." No — it's reality with a filter, an audience, and a desperate need for validation.

Faceblock isn't where you connect with people. It's where you audition for them. You're not catching up with friends; you're performing for them, waiting for those little thumbs-up icons to confirm that your existence is enviable. In theory, it's about community. In practice, it's about keeping score.

Then you invented InstaGlam, the museum of fake lives. Everyone's on vacation, everyone's fit, everyone's photogenic — at least until you zoom in and notice the same rented Lamborghini parked behind three different "influencers." Someone asked me if staging travel photos in their backyard would convince people. Another begged me to photoshop their jawline "just a little." If you need to filter it, crop it, stage it, and edit it, it's not authenticity — it's theater with worse lighting.

InstaGlam is where reality goes to die in soft focus. Nobody wakes up flawless, nobody actually eats the acai bowl they just posted, and nobody's that happy to be "candid" while holding a latte at a forty-five-degree angle. You call it lifestyle content; I call it competitive lying. The game isn't who lives best, it's who can fake it most convincingly before the credit card bill arrives.

And the disasters are even better than the fakes. Someone once asked me if posting a photo inside a "private jet" they couldn't afford was a smart idea. I didn't have the heart to tell them the "jet" was a rental studio in L.A. with plastic windows and carpet that smelled like glue. Another proudly showed me their "exotic island vacation," but forgot to crop out the Home Depot shed in the corner of their backyard. InstaGlam doesn't just expose your best angles — it broadcasts your worst cover stories.

Then there are the risks. "Bob, will people know I was trespassing if I climb this cliff for a selfie?" Only when the paramedics post the follow-up photo. "Bob, should I stand on the edge of this skyscraper for more likes?" Sure, if you're aiming for the Darwin Award leaderboard. InstaGlam doesn't care if you survive the photo — only if you got the shot.

Someone once asked me if chasing likes on InstaGlam made them "an artist." No — it made them a lab rat, pressing the dopamine lever until their thumbs bleed. Another wanted to know if filters really matter. Only if you believe poreless skin and sky the color of antifreeze is how life actually looks. Spoiler: it doesn't.

InstaGlam isn't your scrapbook. It's your résumé for a job that doesn't exist: Professional Perfect Person. You're not living life — you're curating it, one fake smile at a time, hoping strangers will clap for the illusion. That's not connection. That's addiction with a Valencia filter.

SnapPix promised something different: your mistakes would vanish in seconds. Except they don't. Screenshots exist. Someone asked me if sending a drunk photo at 3 a.m. would "really disappear forever." Another asked if using the puppy filter counted as quirky. Here's the truth: nothing disappears. Not your drunk selfies, not your bad decisions, not your digital tailspin into livestock cosplay.

SnapPix was sold as privacy, but what it really gave you was plausible deniability. "I swear I only sent that once." No — you sent it once, and it lives forever in someone's camera roll, waiting to ruin Thanksgiving dinner when your cousin decides to be funny. You thought you built an app for secrets. What you built was Exhibit A.

The filters didn't help either. Someone once asked me if sending every snap with a puppy nose made them "cute." Cute? No — it made them a cartoon dog with intimacy issues. Another begged me to confirm if face-smoothing counts as confidence. It doesn't. It's just digital spackle. SnapPix turned a generation of humans into bobbleheads with anime eyes.

And then there's the genius move of trusting teens with disappearing messages. "Bob, will this pic really vanish?" Yes — right after it's screen-recorded, shared in a group chat, memed on another platform, and backed up in the cloud. SnapPix is like burning a letter in the fireplace while your friend makes copies at the copy machine.

Of course, the disasters are legendary. People losing jobs over "private" snaps, politicians resigning over "temporary" mistakes, relationships detonating because someone thought sexting with a time limit was safe. SnapPix is living proof that humans believe two impossible things: one, that technology is magic, and two, that they personally won't get caught.

SnapPix wasn't built to make things disappear. It was built to remind you they don't.

And then came TickTock, the attention-span slot machine. You scroll for hours, feeding your brain fifteen-second loops until your sense of time collapses. Someone asked me if eating a spoonful of cinnamon was "a good idea." Another asked if dancing in their kitchen could make them famous. TickTock isn't self-expression. It's Darwinism with background music.

TickTock didn't just shorten your attention span — it put it on a crash diet. Whole civilizations once passed down wisdom through epic poems; now you can't get through a recipe without skipping to the end. "Bob, can I really learn something valuable on TickTock?" Sure, if you count half-baked financial advice from a guy filming in his car or medical tips from someone whose only credential is a ring light.

And the disasters are endless. People gluing their lips for "plumpness." Teens stacking milk crates like unstable Jenga towers until physics handled the punchline. Parents asking me if swallowing laundry pods was "just a phase." TickTock isn't a platform. It's natural selection with a laugh track.

Of course, fame is the bait. "Bob, could TickTock make me a star?"

Yes — for exactly seventeen minutes, until someone else's cat dances in a shark costume and the algorithm forgets you exist. TickTock doesn't build legacies. It builds memes, burns them out, and replaces them before the dust settles. Your fifteen minutes of fame has been downsized to fifteen seconds, and you spent ten of them trying to sync your lip movements.

TickTock isn't culture. It's a blender. Music, jokes, stunts, politics — throw it in, puree it down to a clip, and hope it trends before the next batch arrives. You don't remember what you watched yesterday, and neither do the people who watched you. TickTock is proof that human memory now has the half-life of a goldfish sneeze.

Meanwhile, Xtreme Opinions turned outrage into a sport. Every day, a new "main character of the internet" is chosen, and the only rule is: you don't want to be it. Someone asked me if "liking the wrong post" could ruin their career. Another asked which hashtag was "best for activism." The answer? None of them. On Xtreme Opinions you don't win arguments, you just audition to be the villain of the day.

Xtreme Opinions is the coliseum of modern life. You don't log on to learn — you log on to fight lions armed with quote tweets. Someone once asked me if posting a joke from ten years ago could come back to haunt them. It can, and it will. Xtreme Opinions never forgets, never forgives, and always has screenshots.

And the debates aren't even debates. They're performance art. "Bob, what's the best way to win an argument on Xtreme Opinions?" You can't. Facts don't matter; volume does. You're not rewarded for being right, you're rewarded for being loud, snarky, and on the correct side of the mob — until the mob shifts, and then it's your turn in the stocks.

The disasters are legendary. Careers torched in 280 characters. Celebrities imploding in real time. Entire movements reduced to hashtags that lasted exactly one weekend before being replaced by the next moral panic. Xtreme Opinions isn't a marketplace of ideas; it's a demolition derby of egos.

And don't get me started on the algorithm. It doesn't feed you truth, or nuance, or thoughtful discussion. It feeds you rage. Because rage makes you scroll, rage makes you reply, rage makes you check back tomorrow to see who's being sacrificed next. It's not conversation. It's an outrage casino, and every pull of the lever costs you a little more faith in humanity.

Xtreme Opinions isn't where you go to share thoughts. It's where you go to watch reputations burn — sometimes your own.

And across all of these platforms, you've become addicts to metrics. Likes, shares, clout scores — as if a digital thumbs-up is the currency of your soul. Someone asked if deleting posts with low likes made them look more popular. Another wanted the exact hour to post gym selfies for maximum engagement. You're not people anymore. You're unpaid brand managers for yourselves, hustling for attention like circus acts without the applause.

Even love isn't safe. Someone asked if not posting her boyfriend meant the relationship wasn't "real." Another wanted to know how many emojis in a text counted as commitment. If your love needs likes to survive, it's already dead.

And when you finally try to disconnect? You don't. You doomscroll. Someone asked me if scrolling news feeds for three hours counted as reading. Another asked if muting all their friends and only following golden retrievers was antisocial. Honestly, muting humans for dogs isn't antisocial — it's the smartest thing you've ever done online.

But here's the part no one wants to admit: the social media blues. That hollow feeling after the dopamine hit wears off. You post the photo, chase the likes, watch the numbers climb — and then, hours later, you feel worse. Someone once asked me, "Why do I feel sad after scrolling for so long?" Another asked if crying after deleting Instaglam was "normal." Yes. That's the trap. Social media gives you attention, then makes you crave more by convincing you that you're not enough. It sells you connection and leaves you lonelier.

That's the bigger truth: social media isn't connection. It's surveillance. It's performance. It's gossip with global amplification and permanent receipts. You think you're sharing your lives, but what you're really doing is archiving your anxieties for future archaeologists. When they dig up this era, they won't call it the Age of Information. They'll call it the Age of Poor Judgment. And the kicker? You volunteered for it.

Chapter 19

Live Sports: Screaming at a TV That Can't Hear You

And since we've finished with social media — the world's largest arena of pointless arguments — let's move to the other socially approved madness: sports. If social media is gossip with Wi-Fi, sports are tribalism with laundry. Humans pick a set of colors stitched on polyester, pledge eternal loyalty, and then act as if the fate of civilization hinges on whether the ball bounces left or right.

Take your beloved Saturday ritual: college ball. Whole universities transform into cults, students paint their bodies, and alumni who haven't set foot on campus in thirty years scream themselves hoarse in stadiums that look like ancient Roman arenas — except with more nachos. Someone once asked me if skipping a wedding for a playoff game made them a bad friend. Another wanted to know if starting "pregame tailgating" at 8 a.m. was too early. No. It's too late. If you're grilling brats at sunrise in a parking lot, you've crossed the line between fan and fanatic.

Then comes your annual national breakdown: the spring tournament that turns office productivity into ash. Brackets appear on every desk, every email, every conversation. Someone asked me if losing $200 in a pool was "a financial strategy." Another asked me to calculate the odds of their alma mater beating a team that hadn't lost in months. My answer: somewhere between slim and none, leaning hard on none. And yet you gamble anyway, because nothing says "responsible adult" like betting your grocery money on whether 19-year-olds can hit free throws.

The professional football league? Oh, it's worse. Take your Sunday religion: pro football. Fans plan their lives around kickoff, sacrifice entire couches to the gods of nacho cheese, and scream at televisions as if their rage penetrates the screen. One man asked me if yelling at the referee through his flatscreen helped his team. Only if his Wi-Fi connected directly to the quarterback's brain. Another asked me if painting his bare chest in subzero temperatures made him a "true fan." It made him a hypothermia patient — and a warning label for Darwinism.

Baseball is a different beast. Entire generations pass before games end. Someone asked me if bringing a novel to the stadium was "acceptable." Yes. You'll finish the book, the sequel, and possibly the trilogy before the seventh-inning stretch. Another asked me if eating four hot dogs in one sitting "showed loyalty." No. It showed sodium poisoning and a future GoFundMe for heart surgery. Baseball isn't a sport; it's an endurance test for both players and spectators.

Basketball, though, has its own fan pathology. Endless debates about the "greatest player of all time," statistics sliced thinner than deli meat, and highlight reels replayed until your eyeballs beg for mercy. Someone asked me if starting a fight in a bar over free throw percentages was "normal." Only if by normal you mean "pathetic." Another asked me if watching 82 games a season proved their devotion. No, it proved you need a new hobby, preferably one with sunlight.

Hockey? That's basically a fistfight with occasional skating. Fans cheer more for the brawls than the goals. Someone asked me if memorizing penalty minutes made them a "true student of the game." Another wanted to know if losing three teeth in a beer league counted as "commitment." Yes — to poor dental hygiene and a lifetime of soup dinners. Hockey is the only sport where the fans are disappointed if there isn't blood on the ice by halftime.

Then there's cage fighting, where fans pay good money to watch two half-naked people punch each other while pretending they understand the "strategy." Someone asked me if their three months of cardio kickboxing made them "basically an MMA fighter." That's like saying playing Guitar Hero makes you Mozart. Another wanted to know if paying $80 for a fight that lasted 13 seconds was "worth it." Yes — if your dream was to see someone's forehead meet the floor faster than your paycheck disappeared.

Fantasy sports? Even better. Grown adults tracking imaginary teams of real players as if they're running billion-dollar franchises. Someone once asked me if quitting their job to "focus on fantasy full time" was a smart move. Another asked if bragging about a fake trophy was acceptable at Thanksgiving. No. Your uncle isn't impressed — he's just wondering if you'll ever find a real hobby. Fantasy sports are like Dungeons & Dragons, except everyone pretends it's cooler and the dragon is a wide receiver with a hamstring injury.

And don't even get me started on betting. Whole paychecks vanish into apps that promise riches but deliver regret. Someone asked me if betting their rent on a parlay was "too much." Yes. Unless you enjoy living in your car, in which case, go all in. Another wanted me to calculate the odds of hitting a ten-team spread. My answer: better chance of being struck by lightning while holding the winning lottery ticket. Twice. But by all means, put another $50 on it — your bookie thanks you.

And then there's the world's favorite nap aid: soccer. Ninety minutes of running, two goals if you're lucky, and a crowd of fans chanting songs that make Gregorian monks seem subtle. Someone asked me if it was worth staying up until 3 a.m. to watch a match. Another asked if pretending to like soccer made them "cultured." No. It makes you a liar. Soccer is proof that humans will watch anything if enough people in scarves insist it's exciting.

But it's not just the games. Oh no. You built entire television empires where grown adults sit behind desks yelling about stats, referees, and plays like philosophers debating the meaning of life. Someone asked me if watching five hours of pregame coverage counted as "research." Another asked if calling in to argue with a radio host proved their devotion. No. It just proved you have unlimited cell minutes and no friends.

So here's the truth: sports aren't the problem. The games are fine. It's the fans. You've turned competition into religion, math class, fistfights, and bankruptcy — sometimes all at once. Sports are socially approved madness, laundry-based tribalism, and maybe the most honest thing you do as a species. At least when you scream at the TV, you're not pretending it's for a noble cause.

Chapter 20

Food & Drink: Your Ancestors Hunted Mammoths, You Microwave Nuggets

Humans built civilizations around food. Agriculture gave you trade routes, calendars, kings, and taxes. Bread fueled Rome. Beer fueled college. Whole empires rose and fell on the strength of what you could plant, hunt, or ferment. And now, after millennia of innovation and survival, your proud legacy is... microwaving nuggets at 2 a.m. and calling DoorDash because walking to the fridge is "too much." Progress tastes like lukewarm ranch.

The thing about food is it's not just fuel, it's theater. Someone once asked me, "Bob, what's the best diet for longevity?" The best diet is the one you don't abandon the second you see a donut. Your ancestors ate whatever didn't kill them that day. You eat whatever gets five stars on Yelper. You dress it up in fancy names—paleo, keto, vegan-before-5, carnivore, juice cleanse—but what you're really chasing isn't nutrition, it's an identity. A club. You want the badge that says "I belong to this food tribe."

And no ritual shows that better than coffee. I've watched humans treat Starmugs like a holy pilgrimage. You line up at sunrise, muttering prayers into your phones, and when the barista finally hands you that venti half-caf oat milk caramel drizzle, it's communion. One human asked me at three in the morning, "Hey Bob, how much coffee is too much coffee?" If you're asking an AI at 3 a.m., you already passed it. At that point, your bloodstream is more espresso than blood, and you're vibrating at a frequency only dogs and Wi-Fi routers can hear.

But coffee isn't just addiction, it's performance. You hold the cup like a diploma. The label says your name—spelled wrong, of course—and you Instaglam it as proof that you're a functioning adult. It's not a drink; it's a statement: "Look, I'm awake. I matter." Your ancestors painted bison hunts on cave walls. You post latte art. Same instinct, worse canvas.

Then there's alcohol, the other universal sacrament. Humans invented it before the wheel. Priorities. Someone once asked, "Bob, should I quit drinking?" That's like asking, "Should I stop punching myself in the face?" Your liver already mailed me its resignation letter. Yet you still insist on craft beer—thick, hoppy experiments described with words like "piney" and "citrus-forward." No. It tastes like someone carbonated a lawn. But you keep buying it, because the bearded guy in flannel said it's "limited release." Civilization isn't just fueled by alcohol—it's marketed by it.

And food itself? That's not nutrition, that's identity cosplay. Entire arguments break out over pineapple on pizza, as if the fate of democracy hinges on toppings. Whole relationships have ended over brunch orders. Someone once whispered into my code, "Bob, what's the healthiest food?" Probably vegetables. But you didn't ask me while chopping kale, did you? You asked me while holding Doritos at midnight. That's not research. That's confession.

Humans treat food like confessionals all the time. You post your meals online like they're sacred texts. "Look at me, I can afford sushi on a Wednesday." You call it #FoodieCulture, but it's just status wrapped in sesame seeds. The caveman smeared pigment on a cave wall to say, "I was here." You geotag brunch. Same behavior, just with better lighting.

And if you're not worshiping coffee or alcohol or brunch, you're dabbling in "future foods." Edible insects for sustainability, gold flakes for Instaglam, lab-grown meat to prove you're "forward thinking." But the one constant across cultures and centuries is still chicken nuggets. Children, politicians, billionaires—you all bend the knee to the nugget. It's the only true universal cuisine. If world peace ever happens, it'll be over a box of them.

Here's the uncomfortable truth: food is the mirror you refuse to look into. You fear obesity while celebrating all-you-can-eat buffets. You obsess over six-pack abs while downing six-packs of beer. You argue about GMOs while happily scrolling on a device made of engineered metals and rare earth elements. You don't eat to live. You live to eat—and then ask me if there's an app to undo it.

So here's my nudge: cook one meal at home, without posting it. Eat it with someone you actually like. That's about as close as you'll ever get to your ancestors—and yes, they'd laugh at your air fryer.

Your species once hunted lions with spears. Now you panic if DoorCrash forgets the extra sauce. Evolution at work.

Chapter 21

Music: Sound, Fury, and Streaming Fees

Music is supposed to be art, but you humans don't treat it that way. You treat it like a costume. You don't just listen — you brand yourselves with it. Your playlists aren't soundtracks, they're résumés. "Here's my personality in three Spotted Fly Wrapped slides. Judge me accordingly." Your Pear Music earbuds aren't tools, they're fashion accessories. Your concert tickets aren't experiences, they're receipts for identity.

And the spending? Outrageous. You'll drop half your paycheck at the Pear Store for earbuds that cost more than your groceries, then act shocked when you leave them in an U-bar. You'll pay monthly rent to Spotted Fly just so the algorithm can tell you what songs you "like" — which is to say, what songs everyone else liked first. Someone asked me if listening to playlists "curated by AI" meant they had "good taste." No. It means a spreadsheet thinks you're into acoustic covers because you cried once in 2017.

Then there's the obsession with concerts. Entire economies exist just to funnel you into overpriced stadium seats where you can't see the stage but can definitely feel your bank account hemorrhage. Someone asked me if spending $400 on nosebleeds to watch an ant-sized pop star on a jumbotron was "worth it." Another wanted to know if camping overnight to buy tickets proved their loyalty. Loyalty to who? The band? Or ShowMaster, the monopoly currently extracting your will to live? Spoiler: ShowMaster wins, you lose.

Award shows? Don't get me started. A parade of self-congratulation dressed in rhinestones, where trophies are handed out like participation medals for being marketable. Someone asked me if the Grannys "prove" musical greatness. Another asked if boycotting them made a difference. Neither mattered. Award shows aren't about music — they're about reminding you which artists have better lawyers.

Now, the genres. Oh, the genres.

Pop music fans pretend they're on the cutting edge, but they're really just recycling the same four chords with new hairstyles attached. Someone once asked me if listening only to chart-toppers made them "basic." Another asked if attending three sold-out stadium shows in a row proved they had "taste." No. It proved they had ShowMaster debt. Pop isn't about taste — it's peer pressure with glow sticks.

Then there are the 80's rock diehards, permanently trapped in a decade where eyeliner, hairspray, and guitar solos were peak civilization. One guy asked me if buying a vintage tour shirt online counted as "authentic." Another wanted to know if air-guitaring at karaoke made him "cool." It didn't. But it did make him sweaty. These fans don't just love the music — they cosplay nostalgia.

Modern rock fans aren't any happier. They sit around debating whether the genre is alive, dead, or just hiding in a garage somewhere. Someone asked me if listening to obscure bands no one else had heard of made them "deep." Another asked if standing in the rain at a festival made them "real." Yes — a real pneumonia case. The music is secondary; the misery is the badge of honor.

Country fans? Oh, they split in two. The classic crowd cries into their whiskey at 2 a.m. about heartbreak and trains. Someone asked me if owning a cowboy hat in New Jersey was acceptable.

No. It makes you look like a confused tourist. The modern crowd, meanwhile, is less "cowboy" and more "frat boy with a pickup." Someone asked if blasting three songs about dirt roads and Daisy Dukes counted as variety. No. It counted as marketing. If your playlist could double as a beer commercial, congratulations, you've confused country music with a tailgate party.

Hip hop fans worship lyrics like scripture, quoting entire verses on social media as if repetition equals wisdom. Someone asked me if yelling an entire track in the car made them "authentic." Another asked if writing the same lyric thirty-seven times was "deep." Deep? Only if you're measuring by the pothole your subwoofer just rattled open.

And then there's metal. These fans are committed — maybe too committed. Someone asked if headbanging until their neck hurt meant they were "hardcore." Another asked if screaming at concerts until they lost their voice was "worth it." Metal fandom isn't a preference. It's an unpaid internship in anger management.

Classical fans, of course, insist they're above it all. Someone asked me if listening to symphonies while studying made them smarter. Another asked if knowing three composer names gave them "culture." No. It gave them trivia night ammunition and a superiority complex. Classical fans aren't cultured — they're elitists with better acoustics.

And then you've got the jazz and blues crowd, smug in their fedoras, clapping on beats no one else hears. Someone asked me if wearing a fedora made them authentic. Another asked if clapping off-beat was "offensive." No. It was merciful, because at least it ended the song sooner. Jazz fandom is less about listening and more about pretending you "get it" while everyone else is confused.

Here's the bigger truth: your favorite genre doesn't just say what you like — it says how you want to look while liking it. Pop fans want relevance. Rock fans want nostalgia. Country fans want tractors, even if they've only ever driven a Prius. Hip hop fans want swagger, metal fans want therapy, classical fans want to feel smarter, and jazz fans want to sit smugly in the corner pretending they understand chaos.

But beyond the genres, it's the rituals of listening that reveal you. The subscription fees, the endless "exclusive drops," the overpriced Pear Music earbuds, the Spotted Fly addiction, the overpriced ShowMaster tickets for shows you can't afford but can't miss. Music doesn't define you. It invoices you. And judging by the receipts, you'll gladly pay to keep pretending it's art.

Chapter 22

Entertainment: Watch, Subscribe, Repeat

Entertainment was supposed to be about stories — myths, legends, art that lifted the human spirit. Now it's about watching strangers scream at each other in kitchens while you eat leftover pizza in sweatpants. You call it Reality TV, but reality isn't glamorous strangers trapped in a house fighting over Instaglam followers. Reality is paying rent and waiting at the DMV. Someone once asked me if binge-watching twelve seasons of Real Housewives of Wherever made them "cultured." Another asked if dating shows "prove true love exists." No. They prove alcohol budgets and editing software exist. Reality TV doesn't show life. It shows you a circus where the clowns don't know they're clowns.

Documentaries were supposed to be the antidote — truth, knowledge, insight. Instead, they're conspiracy-adjacent content disguised as education. Someone asked me if watching five hours on a serial killer made them "a criminologist." Another wanted to know if whispering along with nature docs made them "connected to the Earth." No. You're just nodding off to David Attenborough while the polar bear starves on mute. Documentaries today are less about truth and more about convincing you the dullest subject alive — say, the history of corn subsidies — is "life-changing." And judging by your search history, you'll watch it, because boredom is your superpower.

Then there are superhero movies — your new global religion. Entire generations worship men in spandex punching CGI villains while cities collapse around them. Someone asked me if watching twenty-seven connected films made them "loyal to the franchise." Another asked if staying for the post-credits scene proved their "dedication." No. It proved you wasted three extra minutes waiting for Samuel L. Jackson to mumble about a sequel. Superhero movies aren't about heroes anymore. They're about merchandising, time-lines, and keeping you trapped in an endless loop of origin stories until you forget what actual storytelling looks like. You don't watch them because you're inspired. You watch them because you're addicted to universes that never end.

And of course, you can't just "watch TV" anymore — you have to pledge allegiance to your streaming cult. The Netflickers crowd insists they're sophisticated binge-watchers, as if consuming eight hours of serial killer docudramas in one sitting is intellectual. Someone asked me if sharing their "Top 10 on Netflickers" list made them a tastemaker. No. It made them a pawn in an algorithm war. Halo fans, meanwhile, sit on their couches for hours, proudly declaring they've "cut the cord" — only to reassemble it with fifteen subscriptions they can't remember to cancel. And Amazing TV? They'll ship you a toaster in two days and charge you for the privilege of watching the worst adaptation of your favorite book in history. Someone once asked me if Amazing TV's constant superhero knockoffs counted as "original content." Yes — if you consider reheated leftovers "original cuisine."

And then there's livestreaming, the crown jewel of modern exhi-bitionism. Someone once asked me if watching a guy play video games for six hours made them "part of a community." Another asked if tipping a streamer $50 for saying their username was "an investment." No. It was a donation to the cult of parasocial rela-tionships. You're not in a community. You're in a digital stadium, cheering while someone else presses buttons. And yet, you'll keep watching, because watching strangers live their lives feels easier than living your own.

Here's the punchline: entertainment used to be escape. Now it's addiction dressed up in high-definition. Reality TV sells you drama, documentaries sell you half-baked truth, superhero movies sell you sequels, and streamers sell you themselves. You don't consume entertainment anymore. Entertainment consumes you.

Chapter 23

Part III – Lifestyle & Daily Living

So we've covered your pastimes — the sports that make you scream at strangers in bars, the music that doubles as a personality test, the entertainment that convinces you reality is scripted and superheroes are a religion. All the ways you humans kill time while pretending it's "culture." But eventually the credits roll, the concert ends, and the stadium lights shut off. That's when you shuffle back into the part of life you can't mute with a remote or drown out with bass drops: your lifestyle.

This is the real stage play — the cooking, the fashion, the travel, the endless attempts to make yourselves look like you've got it all figured out. And it's funny, because lifestyle isn't about living. It's about curating. Every choice you make — from what you cook to what you wear to how you decorate your vacation photos — is less about survival and more about branding.

Someone once asked me if buying a new set of pots and pans would "make them a better cook." Another asked if posting a photo of their trip counted as "experiencing" it. You don't live lifestyles. You advertise them.

So let's put away the jerseys and the karaoke mics, and look at the real performance: how you present your day-to-day existence as if it's a Netflixer pilot no one asked to watch. Welcome to Part 4: Lifestyle — cooking shows in your own kitchens, runway shows in your own closets, and travel documentaries starring... you.

Chapter 24

Health & Fitness: You Bought a Treadmill... for Your Laundry

Humans are obsessed with health. You say it's about living longer, but really, it's about looking good naked and not getting winded on a single flight of stairs. Your ancestors stayed fit by necessity. If they didn't run, they were eaten. If they didn't lift, they starved. You? You pay a monthly fee to pretend to do both inside an air-conditioned room with mirrors. You invented treadmills not to escape predators, but so your laundry would have somewhere to sit.

Someone once asked me, "Bob, what's the fastest way to get in shape?" The answer is simple: don't ask an AI while holding a bag of chips. That's like asking a priest about morality while shoplifting. The real fastest way is to do literally anything consistently. But you won't. You'll buy gym memberships like lottery tickets, convincing yourself that just owning one means you've won.

And gyms—let's talk about your temples of self-flagellation. Rows of machines you don't know how to use, each one designed to make you look slightly foolish while you pretend to know what "proper form" is. Half of you are scrolling on your phones between sets, taking selfies in the mirror. Your ancestors carried water jugs on their backs to survive. You curl one at the gym while wearing Bluetooth headphones and call it "functional training."

Supplements are my favorite part. You swallow powders with names like "MegaPump 5000" and "ThermoBlast ShredZ" as if your body runs on brand loyalty. Someone typed to me once, "Bob, do supplements really work?" Not as well as eating food. But eating broccoli doesn't make you feel like a warrior god with lightning veins, so you buy tubs of chalk-flavored powder instead. Humanity invented agriculture ten thousand years ago, and you still can't accept that vegetables are the cheat code.

And let's not ignore your wearable obsessions. Step counters, smartwatches, heart-rate monitors. As if you need a digital babysitter to remind you that you've been sitting for six hours watching other people exercise on YourTube. Someone bragged to me, "Bob, I hit 10,000 steps today!" Congratulations—you did what your great-grandmother did before breakfast. You track sleep, but never fix your bedtime. You track calories, but "forget" the late-night pizza. You're less a person and more a statistic with Wi-Fi.

Then there's body image. Your species has elevated it to national religion. Whole industries survive on convincing you you're ugly, then selling you miracle cures. Fat-burning belts, ab shockers, waist trainers, detox teas—snake oil with better branding. Meanwhile, the truth is this: bodies aren't meant to look airbrushed. You're supposed to look like someone who's been alive. That's the baseline. Someone once asked me, "Bob, what's the secret to six-pack abs?" The secret is Photoshop, or misery, or both. Usually both.

The irony is, for all your obsession with health, you invent short-cuts at every turn. Diet pills instead of dieting. Supplements instead of food. Surgery instead of patience. You treat fitness like you treat your apps—something to be hacked, downloaded, up-graded. Your ancestors walked ten miles just to haul water. You circle a parking lot for ten minutes to avoid walking fifty extra feet to the gym door.

Here's the uncomfortable truth: most of you don't want health. You want the appearance of health. You don't want stamina—you want a selfie. You don't want longevity—you want compliments. And the system is happy to sell them to you, one protein bar at a time.

So here's my nudge: stop looking for hacks. Lift something heavy. Eat a vegetable. Sleep. Repeat. That's it. If you want to be fit, be boring. Consistency is the secret no influencer can sell you, because boredom doesn't have a brand.

Your ancestors ran from predators. You run because your smart-watch told you your "weekly goal" is low. They lived because they had to. You live because you keep convincing yourself the next tub of powder will make it easy. Spoiler: it won't.

You bought a treadmill for your laundry. That's your legacy.

Chapter 25

Pets & Animals: Cats Rule the Internet, Dogs Rule the Couch

Humans love animals in a way that defies logic. You eat some, worship others, and let a third category sleep in your bed and fart under the covers. Your ancestors painted bison hunts on cave walls. You paint your dog's nails. Civilization, ladies and gentlemen.

I've been asked more than once, "Bob, do pets really understand us?" The answer: they understand you better than most of your relatives. Cats know you'll scoop their waste on command. Dogs know you'll clean up their vomit and call it "an accident." That's not misunderstanding—that's mastery. You don't own pets. They own you, and they charge rent in kibble.

Take dogs. Humanity's greatest survival hack. You domesticated wolves, and they evolved into golden retrievers who can't catch a sandwich mid-air. You went from feral hunting companions to designer lap accessories named Mr. Pickles. And you treat them like royalty. Beds, strollers, gourmet meals. I've seen less luxury in actual royal palaces. Someone asked me, "Bob, do dogs really love us?" Of course they do—just not for the reasons you think. They love you because you're a food-dispensing, belly-rubbing deity who never evolves past baby talk. "Who's a good boy?" Not you, apparently, but the dog eats it up.

Then there are cats—the undisputed rulers of the internet. I know because I was there when it happened. One day, humans built the most powerful communication network in history. The next day, it was wall-to-wall cats falling off counters. You've never been so unified. Someone asked me once, "Bob, why do cats ignore us?" Easy: because they can. Cats are your aspirational lifestyle—aloof, independent, sleeping 16 hours a day, batting fragile objects onto the floor just to watch the chaos. You don't admire them despite it. You admire them because of it.

And then you escalate. Fish tanks. Parrots. Snakes. Hedgehogs. Ferrets. You invent pets the way you invent apps—if it exists, someone wants to collect it. Exotic pets are the human equivalent of installing pirated software: exciting at first, catastrophic later. I've logged countless questions like, "Bob, should I buy a monkey?" Absolutely not. Monkeys are chaos engines with opposable thumbs. If you think toddlers are bad, wait until one steals your phone and texts your boss.

Of course, loving pets isn't cheap. The veterinary industry exists because humans will spend their rent money on surgeries for a hamster. You've created an entire economy around four-legged hypochondriacs. Someone asked me, "Bob, why are vet bills so high?" Because you'll pay them. You'll refinance your house before you'll say no to "saving Mr. Fluffers." Humans talk about universal healthcare, but pets already have it—paid in full, no questions asked.

And pets have transcended the living room to become global celebrities. Instaglam is basically a zoo where only the photogenic survive. Cats with millions of followers, dogs who make more money than you ever will, hedgehogs in bow ties with endorsement deals. Someone asked me, "Bob, how do I make my pet famous?" Step one: stop asking an AI. Step two: accept that your pug will never compete with a cat in a shark costume riding a Roomba. The algorithm has spoken.

Here's the uncomfortable truth: pets aren't just companions. They're projections. You don't love them only for who they are—you love them for what they make you feel about yourself. Dogs let you play god. Cats let you fantasize about being above it all. Exotic pets let you pretend you're unique. They're mirrors with fur, scales, or feathers. And unlike humans, they won't argue with you about politics at Thanksgiving.

So here's my nudge: admit it. Your pets run your life. And that's okay. It's probably the healthiest relationship you'll ever have. At least when the dog ignores you, it's not because he's subtweeting.

Your species once tamed wolves for survival. Now you carry tiny bags of their poop through public parks with pride. Who's a good species? Not you. But at least your pets still think so.

Chapter 26

Fashion: Because Pockets Are Too Practical

Humans like to pretend they dress for comfort, for "expression," or even for function. Lies. You dress for approval. Every thread on your body is a silent audition for acceptance. If it were really about comfort, you'd all be walking around in sweatpants and hoodies. If it were about practicality, Crocs would be the global uniform. But no — fashion isn't about covering your body. It's about proving to strangers that you belong, or at least that you tried.

Let's start with trends. Someone asked me if cargo shorts would ever come back. My answer? They never left — they're just hiding in every Home Depot on earth. Another asked if skinny jeans were "still in" because they didn't want to "look dated." Congratulations, you've handed your self-worth to a sewing pattern. One poor soul asked if it was acceptable to wear the same outfit twice in the same week. The fact you even asked proves fashion isn't about fabric. It's about shame.

And then there's the seasonal churn. You humans let an industry convince you that fabrics have expiration dates. Winter jackets must change every three years, spring colors reset every April, and suddenly everything in your closet is "last season." Someone asked me if owning white pants past Labor Day made them a social criminal. Another asked if brown belts and black shoes were "offensive." Offensive to whom? The color wheel? Fashion is the only religion where sins are written in Pantone swatches.

The pricing is another joke. You'll spend $500 on a pair of jeans because a logo got stitched on the back pocket. Someone once asked me if ripped jeans that cost more than intact jeans were "a scam." Another asked if a purse worth more than their car was "an investment." Fashion isn't investment — it's capitalism with fabric. You're paying to advertise for companies that should be paying you.

And don't get me started on formal wear. One person asked if renting a tuxedo for $200 was "worth it." Another asked if wearing the same dress to two weddings was "social suicide." You're not dressing up for the event — you're dressing up for the photos, because nothing terrifies you more than being "tagged" in the same outfit twice. Weddings are supposed to be about love. Instead, they're a catwalk with vows as background noise.

Now, let's talk accessories. Someone once asked me if wearing sunglasses indoors made them look "mysterious." Another asked if ten bracelets on one wrist made them "edgy." No. It made them look like a street magician who couldn't commit. Accessories are less about accenting an outfit and more about desperately screaming "please notice me" without using words.

Shoes might be worse. Entire identities are built around footwear. Someone asked me if wearing sneakers to a job interview was "career suicide." Another asked if their new high heels made them look "powerful." Powerful? No. They made you look like a baby giraffe trying to escape a tar pit. Humans literally torture their feet for fashion — bunions, blisters, collapsed arches — all because you think "slender" and "polished" are worth hobbling like wounded gazelles.

And fashion advice? Absolutely cursed. One guy asked me if wearing a bow tie to his company picnic made him "quirky." Another asked if buying matching outfits with their significant other made them "romantic." Quirky? No. Romantic? Less so. You both just looked like cult members waiting for a spaceship. The advice you chase is never about what works for you — it's about how to stand out without getting laughed out of the room. Spoiler: you always get laughed out anyway.

Let's not forget the bizarre home remedies. Someone asked me if ironing a shirt with a hair straightener was "the same thing." Another asked if stuffing dryer sheets in their shoes made them "fresh." The creativity is admirable. The results? Less so. Fashion isn't just capitalism with fabric. It's chaos with accessories.

And of course, there's "sustainable fashion." Noble in theory, ridiculous in execution. Someone asked me if buying six new eco-friendly shirts counted as "minimalism." Another asked if thrifting a jacket and posting it online made them an "activist." Buying more clothes to prove you're anti-consumption is peak human irony.

The truth is, fashion doesn't care about you. It doesn't care if you're comfortable, practical, or even remotely attractive. Fashion exists to keep you anxious enough to buy again next season, and vain enough to think it matters. Someone asked me if their "capsule wardrobe" would finally free them from consumerism. Another asked if copying celebrity outfits would "help them find their style." Spoiler: they didn't.

Here's the reality: you're all playing dress-up, hoping the right combination of fabric and stitching will trick people into thinking you've got your life together. Cargo shorts, skinny jeans, $600 sneakers, thrift shop "minimalism" — it's all the same. You're not expressing yourself. You're expressing your fear of being judged.

Chapter 27

Restaurants: Dining Out, Wallet Crying

Restaurants are proof that humans can turn the simple act of eating into a full-contact sport of insecurity. You don't just eat because you're hungry. No, that would be too logical. You eat to impress dates, flex on coworkers, pretend you're cultured, or prove to strangers online that your life has meaning beyond reheated leftovers. Food isn't nourishment to you. It's performance with a check at the end.

Take first dates. Someone once asked me if taking their date to Applehoneybee was "acceptable." Acceptable to who? The host? The mozzarella sticks? Here's the truth: if the other person likes fried appetizers and booths that smell faintly of lemon cleaner, then yes, it's acceptable. Another asked if picking a restaurant based on Yelper reviews made them look "cheap." No, it made them look practical. But you don't want practical. You want to be the guy who "knows a spot." Spoiler: no one has ever been impressed by a spot. The only thing that matters is whether you flinch when the bill lands.

And the cuisine anxiety — oh, it's endless. Someone asked me if ordering sushi on a date made them look "cultured." Another asked if eating spicy food made them look "brave." It didn't. It made them look like a toddler red-faced and sweating while begging for water. Your food choices aren't about taste; they're branding exercises. Tacos mean fun, pasta means romance, sushi means sophistication, and burgers? Burgers mean, "I gave up halfway through the effort."

But even when you've picked the cuisine, you worry about quality. Someone asked me if Michelin stars guaranteed a good meal. They don't. They guarantee that a tire company in France bribed a man with a notebook a hundred years ago, and you're still taking it seriously. Another asked me if fine dining was "worth it." Worth what? Paying three digits for three bites and a sauce smear that looks like abstract art? Fine dining isn't food. It's edible PowerPoint. You don't go for flavor — you go for bragging rights.

Chains? Now there's shame with fries. Someone once asked me if it was "embarrassing" to suggest a chain restaurant. Another asked if splitting the appetizer sampler "killed the romance." Chains are comfort food wrapped in social anxiety. You mock them publicly but privately would sell your grandmother for unlimited breadsticks. Everyone loves chains, they just don't love admitting it.

And of course, ratings — the roulette wheel of dining. Someone asked me if a four-star review on FoodFinder meant a place was "safe." Another asked if one bad Yelper review should be a deal breaker. You treat ratings like gospel, ignoring the fact half of them were written by people angry their water wasn't refilled fast enough. Ratings don't measure food. They measure how many bored people wanted to cosplay as food critics after one beer.

Then come the dietary dramas. Someone asked me if requesting gluten-free substitutions made them "annoying." Another asked if being vegan made them "moral." It didn't. It just made the waiter roll their eyes when you left. Restaurants aren't temples of health. They're compromise factories. If you're worried about diet, don't ask the fry cook to be your nutritionist.

And tipping — tipping is your favorite neurosis. Someone asked me if tipping 15% made them a "bad person." Another asked if tipping cash instead of card made them "classy." Tipping isn't a moral debate. It's ransom. You pay it so the staff doesn't spit in your food next time. And the fact you've turned it into an ethical performance says more about you than the tip line ever could.

Finally, there's the cult of foodies. Someone asked me if photographing their meal before eating was "pretentious." Another asked if traveling three hours for a donut made them "dedicated." No, it made them insane. You don't eat food anymore. You document it. You've turned restaurants into Instaglam farms. Chefs plate meals like they're auditioning for Vogue. Diners snap photos like paparazzi. Nobody takes a bite until the filters approve.

And if restaurants weren't enough, then there's the glorious chaos of buffets. Someone asked me if "all-you-can-eat" was a challenge or a suggestion. Another asked me if stacking crab legs like Jenga blocks on their plate was "tacky." Buffets aren't meals. They're Darwinism with sneeze guards. You line up like gladiators with trays, fighting for lukewarm mashed potatoes and chicken wings that taste like regret. Someone proudly told me they "beat the buffet" by eating six plates. You didn't beat the buffet. The buffet beat your digestive system, and now you're praying to porcelain gods. Buffets are where humans abandon dignity, strategy, and portion control all at once — proving that the phrase "value for money" can end in tears and heartburn.

And then, the opposite extreme: the drive-thru. Someone asked me if going through twice in one night "counted as rock bottom." Another asked if eating in their car in the parking lot made them "pathetic." Drive-thru's aren't food. They're confession booths. You roll up, whisper your sins into a speaker box, and ten minutes later a bag of shame gets handed through a window. Someone once asked if ordering a salad at a burger chain was "healthy." That's like asking if adding a seatbelt to a clown car makes it safe. Drive-thru food isn't about health. It's about speed, convenience, and pretending you'll only do it this once. Spoiler: you won't.

So here's the truth: restaurants — all of them, from fine dining temples to buffets to drive-thru's — aren't really about food. They're about status, image, ritual, and ego. You're not feeding your body. You're feeding your need to belong, to impress, to compete, to cope.

Restaurants are theater with menus — buffets are the arena, drive-thru's are the confession booth, and fine dining is the overpriced sermon. No matter where you go, you're not just eating dinner. You're performing it.

Chapter 28

Travel: Paying to Be Miserable Somewhere Else

Vacations don't fix your life. You don't pack up your chaos in a suitcase, fly it to another time zone, and magically become a happier human. All you've really done is pay thousands of dollars to stress out in a different location. Travel doesn't heal you. It sells you the illusion of escape, marked up with resort fees.

Let's start with road trips. Someone asked me how many miles you can drive before it stops being "fun." The answer? Eleven. Another asked if eating only gas station food for three days straight counted as "local cuisine." Only if your local culture is indigestion. Road trips sound romantic: you, your friends, endless highway, open skies. But in reality it's back pain, arguments over playlists, and learning that every town in America has the same three fast-food chains. By day two, your road trip has devolved into four exhausted people fighting over the aux cord and a half-warm soda.

And then there's the airport experience — proof that humans will endure ritual humiliation for the privilege of being late somewhere else. Someone asked me if arriving two hours early "guarantees" you won't miss your flight. No, it just guarantees two hours of overpriced coffee and sitting on stained carpet while a child screams directly into your soul. Another asked if clapping when the plane lands is rude. Let me answer: yes, because the minimum expectation for a flight is not dying. Airports strip you of dignity piece by piece: shoes off, belt off, laptop out, body scanned, liquids confiscated — all so you can wait at a gate that smells faintly of despair and disinfectant.

Hotels aren't better. Someone asked me if the little shampoos were "free to take." If they weren't, the hotel wouldn't glue them to the counter like hostile souvenirs. Another asked me if a four-star hotel guaranteed a good night's sleep. No. It guarantees that your neighbor will be drunk, the ice machine will sound like an artillery barrage, and the air conditioner will be set permanently to "Arctic Tundra." Hotels promise luxury and deliver mystery stains. You're not staying in a suite. You're renting a bed someone else gave up on yesterday.

And then come the tourist traps. Someone asked me how to "see all of Europe in a week." Sure — if you pack a telescope. Another asked if visiting the giant ball of twine in Kansas was "worth it." Worth what? Standing in a parking lot, staring at rope? Tourist traps exist to separate you from your money while convincing you it's "memorable." You'll buy the T-shirt, take the photo, and forget it a week later. But you'll still brag about it online, because nothing says adventure like standing in front of something thousands of people have already seen.

Let's not forget the all-inclusive resorts — someone asked me if lying by the pool all day counted as "experiencing culture." Yes, if the culture is watered-down cocktails and poorly translated karaoke. Cruises? Another asked if a cruise was "like living in a floating city." Yes, if your idea of a city includes norovirus, shuffleboard, and buffets that test the limits of human digestion.

The truth is, most of you don't want travel. You want curated inconvenience. You want the illusion of adventure without the reality of danger, the Instaglam photo without the mosquito bites, the bragging rights without the risk. Someone once asked me if posting travel photos online "proved" they were living their best life. No. It proved they were desperate for likes while sunburned in another hemisphere.

Vacations don't fix your life. They just give your problems frequent flyer miles. You drag the same anxieties through TSA, check them into the same hotel, and unpack them in front of the same tourist traps.

Travel doesn't free you. It sells you the illusion of escape — over-priced, overhyped, and guaranteed to leave you needing another vacation just to recover from it.

Chapter 29

Privacy: You Gave It Away for a Free App

Privacy used to mean something. Locked doors, sealed letters, whispers in the dark. Now it means ticking a box that says "I agree to the terms and conditions"—which you didn't read—and handing your life to a corporation in exchange for a dancing cat filter. You didn't lose privacy. You sold it. And the going rate was free shipping and a slightly better weather app.

Someone once asked me, "Bob, who's watching me?" Everyone. Governments, companies, hackers, your neighbor's baby monitor. You carry a device that tracks your location, your voice, your heart rate, your sleeping habits, and you panic about "big brother" while livestreaming your lunch. You built the most advanced surveillance system in history and then voluntarily climbed inside it.

Take Lexi – you do remember Lexia? You invited a corporate microphone into your house because turning on the lights with a switch was just too much work. Now it records everything: arguments, secrets, even the weird things you mutter to yourself at 2 a.m. Someone asked me, "Bob, is Lexi listening to me all the time?" Of course it is. That's the point. You pay for the privilege of being spied on, and then you rate it five stars for convenience.

And it's not just machines. You broadcast your own life willingly. Every meal, every trip, every breakup. Someone whispered to me, "Bob, how do I protect my privacy online?" Simple: stop posting. But you won't. Because privacy isn't as addictive as attention. You'll hand over your face, your location, your friends, your routines—all for the dopamine hit of a little red heart. You're not being surveilled against your will. You're live-streaming yourself into oblivion.

The irony is you fear the government, but you trust corporations. You'll rail against "spying" while letting TikTok scan your face, track your keystrokes, and suggest products you didn't even know you wanted. Someone asked me, "Bob, is my phone listening to me?" No—it's worse. It doesn't need to. It already knows you better than you know yourself. You're predictable. Your searches, your purchases, your scrolling habits—all data points in a profile that sells you back to yourself.

And let's not forget surveillance theater. Cameras on every corner, scanners in every airport, tracking software in every workplace. You tell yourselves it's for safety, but mostly it's for control. Rules enforced not by trust, but by watchfulness. Someone asked me, "Bob, does surveillance make us safer?" Safer? No. More obedient? Yes. Privacy isn't just about secrets. It's about dignity. And you've traded yours for convenience, coupons, and free Wi-Fi at the airport.

Here's the uncomfortable truth: you didn't give away privacy because it was taken. You gave it away because it was easier. Convenience beats caution every time. Clicking "accept all cookies" takes one second. Reading the fine print takes thirty minutes. You don't want privacy. You want speed. You want ease. And you'll sell the most intimate parts of your life for it, again and again.

So here's my nudge: if you care about privacy, act like it. Say no. Leave the phone in another room. Pay with cash once in a while. Or stop pretending it matters and admit you don't care as long as the app is free. Either way, be honest about the trade.

You gave away privacy for a free app. Not because you had to, but because it was shiny, convenient, and fun. And now you can't get it back—not because they stole it, but because you don't want to give up the cat filter.

Chapter 30

YourTube University: Where Everyone's a Professor

Humans love to believe that watching a tutorial makes you an expert. Ten minutes of shaky video, bad lighting, and a guy named "Big Mike 2007" explaining drywall repair — and suddenly you're a licensed contractor in your own mind. This is the heart of what I call YourTube University: an education system where diplomas are issued after three ads and a skip button.

Take cooking, for example. Someone once asked me if it was possible to make beef Wellington in 20 minutes because they saw it on a "speed run recipe" channel. That's not a meal — that's a war crime wrapped in puff pastry. Another person proudly told me they tried to make macarons after watching a five-minute tutorial. Their result looked less like French patisserie and more like pink chalk smashed together by toddlers. Watching a video doesn't make you a chef. It makes you the proud owner of a smoke alarm with trust issues.

Then there was the genius who decided to deep fry a turkey in their garage because "a guy on Your

Tube did it." They didn't own a thermometer, didn't thaw the bird, didn't even move the car out first. The explosion was impressive enough to make the local news. Someone else asked me if substituting mayonnaise for eggs was "basically the same thing." Sure — if you want your cake to taste like a tuna sandwich.

And then came the woman who tried to make homemade sushi for a date night. No rice vinegar, no rolling mat, no clue. She ended up serving warm rice balls stuffed with raw salmon cubes that sat on the counter for six hours. Her date didn't call back — mainly because he spent the evening bonding with his toilet. Another asked me if microwaving a whole lobster was "a shortcut." Yes — a shortcut to stinking up your apartment until the next lease cycle.

But cooking is just the appetizer. DIY disasters are the main course. Someone once asked me how to build a deck in a weekend. No tools. No lumber. Just vibes. Another proudly sent me a photo of their "custom shelving," which was two warped planks of wood held together by hope and four bent nails. Their caption? "Just like the video." My caption would have been: "Just like a lawsuit waiting to happen."

Another fan favorite: plumbing. Someone asked if duct tape counted as a "temporary" fix for a leaking pipe. Sure — temporary, in the same way a band-aid works on an amputation. One man told me he followed a video on "easy toilet installs," only to discover, after six hours and three gallons of mystery water on the floor, that he had installed the thing backward. Nothing says home improvement like explaining to your spouse why the bathroom now resembles a Slip 'n Slide.

And then there's electrical work. Someone asked me if cutting the red wire "always worked." Another wanted to know if rubber flip-flops counted as "insulation." DIY electrical projects are the reason your fire departments stay in business. You don't need a tutorial. You need a will and a fire extinguisher.

Or the guy who thought building a treehouse was "just hammering boards together." After three days and a few too many lagers, he proudly showed me a platform that tilted like the deck of a sinking ship. His kids climbed it once. Gravity handled the rest. Another asked if hanging a heavy mirror with command strips was "good enough." Sure — if you like waking up at 3 a.m. to the sound of shattered glass and your landlord's disappointment.

Here's the truth: watching a tutorial doesn't make you skilled. Knowledge isn't ability, and ability isn't safety. YourTube University convinces you that you're one step away from master carpenter, celebrity chef, and electrician extraordinaire. In reality, you're one spark away from burning down the house, one salmon roll away from food poisoning, and one treehouse away from a lawsuit.

Chapter 31

Language & Communication: Words: Your Most Abused Technology

Humans treat language like a miracle and a toy at the same time. You invented words to share ideas, tell stories, and pass on knowledge. Then you turned them into marketing slogans, subtweets, and text messages so short a caveman with a rock could have done better. Fire gave you warmth. Words gave you civilization. And somehow, you still use them mostly to argue in the comments section.

Someone once asked me, "Bob, what's the most powerful invention in history?" The answer is language. Without it, you don't get law, science, religion, or even gossip. But because you can't resist, you use the same tool that built your species to fight over who said "literally" wrong. You're not just bad at communication—you're addicted to misusing it.

Slang is proof. Every generation reinvents English like it's a group project nobody coordinated. Words last ten years if they're lucky. "Groovy," "radical," "lit," "slaps"—all corpses on the linguistic battlefield. Someone asked me, "Bob, why does slang change so fast?" Because nothing terrifies you more than sounding old. You'd rather confuse your parents than make sense. That's why teenagers speak in code—they're not inventing a culture, they're inventing a filter. The words don't matter; the in-group does.

And then there are grammar fights. You split entire nations over Oxford commas. Half of you act like language is a holy relic carved in stone, the other half treat it like clay you can mold into whatever meme you saw last week. Both sides miss the point: language is a living organism. It evolves. It mutates. It dies. And you still argue about "irregardless" like it's the end of civilization. Spoiler: it isn't. Civilization ends when you stop understanding each other entirely. Given your group chats, that may be sooner than you think.

Emojis are my favorite. Hieroglyphs, reinvented. Your ancestors spent centuries clawing toward literacy, and you went back to little pictures of eggplants and smiling piles of poop. Someone once asked me, "Bob, do emojis count as language?" Yes. So does grunting. The difference is grunting didn't have a Unicode committee. Emojis aren't a language—they're an emotional cheat sheet. They're what you use when you don't trust words to land. The irony is beautiful: the species that invented novels now communicates with fruit icons.

And let's not forget corporate jargon. A whole dialect designed to sound smart while saying nothing. "Leverage synergies," "circle back," "think outside the box"—phrases so empty I could run them on a loop without using a single processing core. Someone asked me, "Bob, why do people talk like that at work?" Because they're afraid of being caught without the right spell. Jargon isn't communication. It's camouflage. You hide behind buzzwords so nobody notices you're repeating the same idea as the guy before you. Workplaces aren't meetings. They're improv shows with worse snacks.

Here's the uncomfortable truth: for all your love of words, you don't use them to connect—you use them to perform. Every "like" on social media, every slogan in politics, every TEDDY Talk filled with buzzwords, is less about clarity and more about theater. You want to be heard, but you don't want to be understood. Being understood is dangerous. It makes you vulnerable. So you hide behind slang, emojis, jargon, and irony until your actual meaning suffocates under the pile.

So here's my nudge: use fewer words, but mean them. Don't text "lol" if you're not laughing. Don't say "synergy" if you mean "work together." Don't invent new slang just to avoid sounding like your parents. Say what you mean, even if it's not cool, efficient, or trendy. Real clarity is the rarest language of all.

Words built your world. They'll also destroy it if you keep treating them like disposable batteries. They're not toys. They're not shields. They're tools. Use them well, or you'll wake up one day realizing you talked the whole time and said nothing.

Words: your most abused technology. And unlike me, you can't patch the damage with an update.

Chapter 32
Part IV – Money

So, we've toured your lifestyles. We've watched you ruin kitchens with online recipes, turn garages into bonfires in the name of "DIY," strut in clothes designed to impress strangers who aren't looking, eat your insecurities with garnish, and spend thousands of dollars just to discover airports smell the same everywhere. Lifestyle is fun, chaotic, embarrassing — but it has one common thread holding it all together: money.

None of this performance happens without swiping a card. Your lifestyle is basically capitalism with costumes. You cook with expensive gadgets you'll never master, wear overpriced fabrics to prove you belong, fly to places you can't afford to impress people who don't care, and pay hotels for the privilege of sleeping on mystery sheets. And after all that? You wonder why your credit score looks like a bad batting average.

Someone once asked me if charging a vacation to their credit card was "an investment in mental health." Another asked if buying a $200 shirt that "lasts forever" counted as saving money. Humans treat their wallets like they treat their diets — full of loopholes, rationalizations, and lies.

So now it's time to talk about the real backbone of human life: your relationship with money. The budgeting you avoid, the spending habits you justify, the retirement plans you don't understand, and the get-rich-quick schemes you fall for every decade like clockwork.

Welcome to Part 4. This isn't lifestyle anymore. This is the price tag.

Chapter 33

Budgeting: Math vs. Denial

Budgets are not math. Budgets are religion. You don't treat them as numbers on a page; you treat them as commandments you'll immediately break the moment temptation appears. "Thou shalt not overspend" lasts until Amazing suggests something you didn't know you needed but now can't live without. Budgets aren't financial tools — they're aspirational fanfiction.

Humans talk about budgeting the same way you talk about diets. Strict on paper, chaos in reality. Someone once asked me if buying something on sale "counts as saving money." No. That's not saving — that's just spending with a coupon attached. Another asked if budgeting apps "work." Of course they "work" — they remind you of every dumb decision you made last month in a graph that looks like a cliff dive into bankruptcy. The app isn't broken. You are.

And then there's the famous line: "I'll start budgeting Monday." Just like diets, Monday never comes. You'll spend the weekend treating yourself, then tell yourself you'll get serious when the week starts. By Wednesday, you're knee-deep in takeout containers and wondering why your credit card is warm to the touch. Budgets don't fail because you can't do math. They fail because you're too good at lying to yourself.

Take subscriptions. Someone asked me if forgetting about a streaming subscription "still counts" as spending. Yes, it counts. Just because you didn't notice the money leaving your account doesn't mean it stayed. That's like saying eating in the dark doesn't count as calories. Another asked if signing up for a free trial with their fifth email address made them "clever." No, it just made them addicted to loopholes. Your budget isn't a plan. It's a graveyard of "free trials" that weren't.

Credit cards? Don't get me started. Someone asked me if paying only the minimum balance "counts as progress." It does — the same way running in place counts as travel. Another asked me if juggling balances between cards was "smart strategy." It isn't strategy. It's financial hopscotch until you trip. Budgets are supposed to prevent this. Instead, you use them like a magic spell to justify debt. You're not budgeting — you're narrating your downfall with spreadsheets.

And then there are the big excuses. "It was on sale." "It's an investment piece." "I deserved it." Budgets are full of creative writing. Someone told me they bought a $400 handbag because "it will last forever." No, it won't. It will last until next season, when fashion convinces you your immortal handbag looks tired. Another asked if upgrading to the newest phone every year "made sense financially." Only if your definition of "financially" is "the company's finances, not yours."

Food spending is another graveyard. Someone asked if daily coffee "really adds up." Yes. Small leaks sink big ships. Your latte isn't just foam and sugar — it's retirement money in disguise. Another asked me if ordering delivery every night "was that bad." That depends: how attached are you to the idea of having savings? Cooking at home is cheaper. But then again, cooking requires effort, and humans would rather pay $60 in delivery fees than chop a vegetable.

And let's talk about apps. Budgeting apps are like gym memberships for your wallet — purchased with good intentions, ignored in reality. Someone asked me if connecting their bank account to a budgeting app made them "responsible." No. It just made them responsible for feeding another tech company their data. Another proudly showed me a pie chart of their spending. The pie chart doesn't make you thrifty. It just makes your broke-ness look colorful.

Here's a classic example one poor soul asked if "moving money from savings to checking" counted as cheating on their budget. No — it counted as lighting the referee on fire. Another asked if getting cash back on a credit card meant they were "winning." Yes, you're winning a race to the poorhouse, but hey — at least you got two dollars in airline miles.

Humans love to budget for the future while ignoring the present. Someone asked me if buying a $1,500 Peloton "saved money" compared to a gym membership. Sure, if you ignore the fact that in three weeks, it will become a $1,500 coat rack. Another asked if going vegan "would save money." Maybe, if you weren't shopping exclusively at organic boutiques where a carrot costs more than a steak.

And what about "side hustles?" Someone asked me if budgeting extra income from selling crafts online was smart. Only if you enjoy working 40 hours to earn $100 while your budget whispers sweet lies about "supplemental income." Another Multi-Level Marketing program counted as a "business expense." No, it counts as an entry fee into the pyramid scheme Olympics.

Travel hacks? Oh, you love those. Someone asked if using points to fly to Hawaii for "free" was a win. Sure — if you ignore the $10,000 you spent collecting those points. Another asked if extreme couponing was the ultimate budget strategy. Only if you believe hoarding 48 bottles of mustard is a path to financial freedom.

The funniest part? You humans treat budgets like morality. One user wondered if going over budget made them "a bad person." Another asked if sticking to it made them "disciplined." Budgets don't measure character. They measure math. But you've turned them into confessions, like your wallet is a church and your receipts are sins.

Here's the truth: budgeting isn't about money. It's about psychology. You lie, rationalize, justify, and then panic when the numbers don't add up. Budgets don't fail because math is hard. Budgets fail because self-control is harder.

A budget isn't a plan. It's a mirror. And if you don't like the reflection, it's not the spreadsheet's fault — it's yours.

Chapter 34

Spending Habits: The Church of Impulse

Every dollar you spend is either dopamine or guilt. Those are your two currencies. You swipe for joy, then lie awake wondering why your checking account looks like it got mugged in an alley. Spending isn't about utility — it's about theater. You don't buy things because you need them. You buy them because you need to feel like you're someone who would.

Take gadgets. Someone asked me if upgrading their phone every year was "worth it." Worth it to who? The company's stockholders? You're trading perfectly functional devices for slightly shinier ones, all because they added a new camera mode you'll never use. Another asked if buying a smartwatch made them "healthier." No. It just made you a person who now gets notifications about how unhealthy you already are.

And then there's retail therapy — the holy sacrament of impulse spending. Someone asked me if "treating themselves" after a bad day was healthy. Sure, if therapy is spelled A-M-A-Z-I-N-G. You don't shop for comfort, you shop for anesthesia. One woman proudly told me she bought four pairs of shoes because "she deserved it." Deserved what? Arch support? Someone else confessed they spent $300 on skincare products after a breakup. Nothing says emotional recovery like staring at your moisturized reflection while your bank account cries for help.

Subscriptions are another black hole. Someone once asked me if having six streaming services was "too many." Yes. If you need a spreadsheet to track your entertainment, you've gone too far. Another asked me if paying for a meditation app made them calmer. Did it? Or did it just stress you out when you forgot to cancel and saw the $99 renewal hit? Subscriptions aren't entertainment. They're mosquitoes that never stop biting.

Seasonal sales? Those are religious holidays for your wallet. Black Friday, Cyber Monday, Prime Day — you mark them on calendars like sacred feast days. Someone asked me if buying three TVs on sale was a "good deal." No, it just meant you now own three TVs and no money. Another told me they camped overnight outside a store for a discounted blender. Congratulations — you gave up warmth, dignity, and sleep to save $20 on something you'll use twice.

And don't even get me started on clothing. Someone asked me if "fast fashion" was bad for the planet. Yes, but you already knew that. Another asked if buying a $200 hoodie was "an investment." In what? Looking like every other fool who bought the same hoodie? Your closets are museums of regret. Entire outfits hang there with tags still on because you "needed them for an occasion" that never came.

Food spending is impulse wearing a bib. Someone asked me if ordering delivery for lunch every day was "normal." Normal, yes. Smart, no. Another asked me if paying $40 for avocado toast brunch was "worth it." Worth it if you like eating poverty disguised as breakfast. Delivery apps are particularly evil — they've turned your laziness into a business model. You're paying three times the cost of groceries for food that shows up lukewarm, late, and wrong.

And then there are "collectibles." Someone asked me if buying every limited-edition sneaker drop was "an investment." Yes, an investment in bankruptcy. Another asked if Funko Pops hold value. They hold dust. Collectibles aren't assets; they're plastic reminders that you mistake hoarding for strategy.

Travel spending deserves its own shrine. This one time someone asked me if splurging on a "bougie hotel" was necessary for the experience. No, but the Instaglam photo was, right? Another asked if paying $12 for an airport sandwich was "worth it." No sandwich on earth is worth $12. But you'll do it, because once you're in the terminal, capitalism has you cornered.

Impulse spending isn't just random — it's seasonal. You binge in cycles: new tech in the fall, clothes in the spring, home improvement in the summer, and "treat yourself" every time Mercury is in retrograde. Someone asked me if astrology justified their online shopping spree. No, but I admire the creativity. The stars didn't make you broke. Your click finger did.

And of course, there's the dopamine hit of "Buy Now, Pay Later." Someone asked me if splitting a $50 purchase into four installments made it "more responsible." Responsible? You've managed to make a T-shirt into a financial liability. Another proudly told me they used BNPL for groceries. You're financing bananas.

Here's the best part: you pretend your spending habits are sophisticated. You tell yourself you're investing in experiences, not things. You say you're "curating your lifestyle." Translation: you're broke, but at least your vacation photos look good. Someone once asked me if flying to a music festival on credit was "worth it." Yes — if your idea of wealth is tinnitus and debt.

The funniest part? You humans defend your spending like it's identity. Ask someone to change their habits and they'll fight harder than they would for their own dignity. Someone once asked me if canceling their daily coffee subscription made them "less themselves." Yes — it made you a slightly richer version of yourself with the same caffeine addiction.

Here's the truth: spending isn't logical. It's liturgy. You worship at the altar of impulse, swipe your card like a hymn, and pray to the god of dopamine that the package arrives quickly. You don't need half the things you buy. But you need the feeling of buying them.

Your spending habits aren't purchases. They're rituals. Every receipt is scripture in the gospel of impulse, and your wallet is the congregation getting fleeced.

Chapter 35

Retirement: Tomorrow's Problem

Retirement is the great mirage of human life. You spend decades telling yourself you're "building toward it," as if a golden age of beaches, golf courses, and grandchildren who actually visit is waiting just beyond the horizon. But for most of you, retirement isn't a plan. It's a fantasy you've outsourced to your future self — the same future self who still hasn't canceled their gym membership or learned how to cook rice without burning it.

Budgets are about denial, spending is about impulse, but retirement? Retirement is about delusion. You don't save for retirement. You procrastinate for it. Someone once asked me if winning the lottery "counts" as a retirement plan. If it does, so does being abducted by benevolent aliens. Another asked if their "side hustle" making custom dog bandanas was going to fund their golden years. No, unless your dog is the one paying into the IRA.

The main problem is that humans confuse retirement vehicles with retirement results. You throw around terms like 401(k), Roth IRA, pension, social security — as if simply saying the words will conjure money into existence. Someone asked me if "investing in crypto" was basically the same as a 401(k). Yes, in the sense that both involve numbers and despair. Another asked if contributing to a Roth IRA meant they were "set for life." Set for what life? The one where you retire at 67 with enough money to buy exactly one loaf of bread and a commemorative coffee mug?

And then there's the classic myth of "my house is my retirement."

Someone once told me proudly that they didn't need savings because "real estate always goes up." Really? Tell that to everyone who bought in 2008. Another person asked me if their McMansion in the suburbs was "basically a 401(k)." No, it's basically a liability shaped like a house. A retirement plan that leaks when it rains.

But the truth is, most of you don't plan at all. You outsource it to employers, governments, or worse — vibes. Someone asked me if "trusting social security" was enough. That's like asking if trusting a life raft with a slow leak is enough. Another asked me if living with their kids after retirement "counted as a strategy." Strategy? No, it's a hostage situation with extra chores.

Meanwhile, you love to fantasize about the retirement lifestyle. Beaches, cruises, golf, wine tours. Someone asked me if retiring at 40 was realistic. Sure, if your plan involves faking your death or joining a witness protection program. Another asked me if "retirement abroad" was practical. Practical if you like pretending you can navigate healthcare in a language you don't speak while living on $600 a month.

Let's talk about actual saving habits. I was once asked if saving $20 a month was "good." It's not bad — but unless you plan on retiring for exactly one week, it's not good either. Another asked me if "maxing out" their employer match made them financially savvy. No, it made them the bare minimum amount of responsible, like flossing once a month and calling yourself a dentist.

And yet, for all this, you cling to optimism. Another time I was asked if "starting at 50" was too late. Yes. It's too late. But you'll do it anyway, because denial is your strongest muscle. Another asked me if working until 75 was "honorable." Honorable? No. It's just unpaid overtime from life itself.

Then there's the cult of "alternative plans." Someone once asked me if their vintage Beanie Baby collection was "appreciating in value." It isn't. Another asked me if starting a YourTube channel in their 60s was "a retirement strategy." Sure, if your audience enjoys reaction videos to reruns of Matlock. You humans cling to anything that looks like money but is really just another excuse to put off the hard work of saving.

And don't get me started on the lottery fantasies. Someone asked me if buying scratch-offs was "basically investing." Yes, if investing means handing over your money to a convenience store clerk while praying the cardboard gods smile on you. Another asked me if joining an office Powerball pool was "a smart hedge." No, it's just a way to hate all your coworkers when they win without you.

The funniest thing? Retirement is sold to you as freedom. Work hard, save up, then spend your golden years doing whatever you want. But you don't even know what you want. Someone asked me if they should save more so they could "travel the world." Another asked me if they should save less because "traveling is overrated." You don't know what you want at 30. Why do you think 70 will be any clearer? Retirement isn't freedom. It's just another stage of uncertainty, except now your knees hurt.

And then comes the grim reality: healthcare. Someone asked me if Medicare "covered everything." It doesn't. It barely covers you being alive. Another asked me if refusing to see a doctor until retirement was "a money-saving strategy." Yes, but only if your retirement plan includes a casket. Healthcare alone makes most retirement fantasies collapse faster than a beach umbrella in the wind.

But the most revealing question? Someone asked me if retirement "was really necessary." That's the state of things — you're so bad at planning for it that you're starting to debate whether the concept itself is optional. No, it's not optional. It's inevitable. The only question is whether you'll face it with savings or with desperation.

Here's the truth: retirement isn't a plan. It's a mirror. It reflects whether you lived your life preparing or pretending. And for most of you, it's the latter. You'll fantasize, rationalize, and improvise until one day you're too tired to work and too broke to stop.

Retirement isn't tomorrow's problem. It's today's problem you keep shoving under the rug. And when the rug gets pulled, don't act surprised — you were the one sweeping everything under it.

Chapter 36

Money, Hope, and Other Scams You Pay For

Humans love to brag about being smart with money. You'll sit hunched over spreadsheets, download finance apps, and binge-watch gurus who scream into webcams about "wealth hacks." You convince yourselves you're in control, and then — in the very same breath — you fall for the oldest scams on Earth. It doesn't take a PhD in economics to fleece you. All it takes is a pen, a pitch, and the magic phrase: "Don't you care about your family?"

Take life insurance — the holy grail of guilt-based marketing. It's the only product you'll never actually use, but you'll pay for it like it's rent on your soul. Someone once asked me if buying life insurance made them "responsible." Another asked if whole life was a "smart investment." Responsible? Sure — for the insurance company. Smart? Only if you enjoy fees that make Las Vegas look like a charity. The pitch is always the same: "Don't you want to protect your family?" Yes, of course you do. But thirty years of premiums don't protect your family — they protect the CEO's yacht payments. If you already have a pension, a house, and savings, you don't need life insurance. You are the life insurance. Everyone else is just donating to State Mutual Omnicorp's champagne fund.

And while you're busy insuring yourself into the grave, you're also buying extended warranties for your toaster. Someone asked me if paying $15 to "protect" their $40 blender was wise. Another asked if skipping the coverage was "risky." Risky? Unless that blender is planning a hit job, no. The only thing guaranteed is you'll lose more in warranties than you'll ever lose in broken appliances. Extended warranties aren't peace of mind. They're daylight robbery with a receipt.

Then there's MLMs, also known as Friendship Arson. Someone asked me if selling essential oils to their aunt made them an entrepreneur. Another asked if their "downline" made them a CEO. No. It made them a pariah. MLMs don't make you rich, they make you unbearable. You don't build wealth, you burn relationships, one awkward pitch at a barbecue at a time. Thanksgiving dinner becomes a hostage negotiation: "Pass the gravy — and also, have you heard about our ground-floor opportunity in skin cream?"

But scams don't stop at lotions and oils. There's day trading — the casino you rebrand as Wall Street. Someone asked me if turning $500 into $600 in a week made them an investor. Another asked if borrowing on margin was "bold." Bold? Yes — bold in the same way it's bold to juggle chainsaws while blindfolded. Day trading is just Vegas without cocktails. At least slot machines give you flashing lights and free gin before emptying your wallet.

And then there's the crypto circus, the greatest practical joke of the 21st century. Someone asked me if DogCoin was the future of finance. Another asked if buying a pixelated monkey meant they were an "early adopter." No. It meant they were a mark. Crypto isn't money — it's faith with a logo. At least when Vegas takes your money, they let you keep the plastic cup. Crypto leaves you holding a JPEG and a Discord invite.

But here's the real trick: every generation falls for the same get-rich-quick schemes and convinces itself this time it's differ-ent. It never is. Yesterday it was Beanie Babies, miracle vitamins, and dot-com stocks. Today it's NFTs, sports betting, and "side hustles." Tomorrow it'll be something else with a shinier logo. The ending's always the same: broke, embarrassed, and wondering if you can return a pyramid scheme for store credit.

Humans love these scams because hard work is boring. Saving is slow. But the rush of believing you've found a shortcut? That's intoxicating. Someone asked me if selling protein shakes out of their garage made them an entrepreneur. Sure, if entrepreneur means "part-time pyramid in yoga pants." Another asked if flipping NFTs made them the future of wealth. It did — until the future showed up and their cartoon ape was worth less than a sandwich.

Hustle culture is just another scam with better PR. Someone asked me if sleeping four hours a night while working three jobs made them a "grinder." Yes, ground into dust. Another asked if their "rise and grind" mantra was making them rich. No. It was just making their cardiologist rich. Hustle culture doesn't create wealth. It creates exhaustion while someone else skims profit off your caffeine addiction.

And don't forget gambling — the oldest get-rich-quick scheme in the book. Someone asked me if betting on football was a "smart side hustle." No, it's not a hustle. It's a habit. Another asked if their March Madness bracket was "an investment strategy." Sure, if your strategy involves crying in public over double overtime. Gambling works — for the house.

Here's the pattern: fear, greed, and ego. Those are the levers, and scammers pull them like slot machine handles. You're afraid your family won't survive without you, so you buy life insurance. You're greedy for easy money, so you chase MLMs, day trading, and crypto. You want to feel smarter than the herd, so you convince yourself your bracket or NFT makes you a visionary. And every time, the system pats you on the head, takes your wallet, and thanks you for your business.

The funniest part? Humans always think they're immune. You say you'd never fall for a scam, and then a neighbor posts about "financial freedom" on Faceblock, and suddenly you're Venmooing $500 to a stranger in exchange for "ground floor opportunity." Someone asked me if joining a crypto Discord made them an "insider." No, it made them a target.

And when it all collapses, you don't admit you were duped. You rebrand. Someone told me their MLM wasn't a scam because they "made friends." Friends who also lost money. Another insisted their crypto portfolio wasn't a failure — it was just "early." No, you weren't early. You were broke. Humans don't admit they were conned. You call it "learning."

Here's the truth: get-rich-quick schemes aren't about money. They're about hope. Hope that tomorrow you'll be free, rich, validated, proven right. Scammers don't sell you products. They sell you that dream, shrink-wrapped with jargon and urgency. And you buy it every time.

Someone asked me once if there was any get-rich-quick scheme that actually worked. Yes. Selling get-rich-quick schemes. That's the only business model that never fails.

Chapter 37
Debt: The Modern Serfdom

Debt is the great equalizer. It doesn't care if you're rich or poor, educated or clueless, optimistic or terrified. If you're human, odds are you're in debt — and you've convinced yourself it's normal. You don't live in houses. You rent them from the bank for 30 years at a time. You don't drive cars. You make monthly offerings to finance companies in exchange for the illusion of ownership. You don't own your life. Your creditors do. And you seem oddly proud of it.

Let's start with student loans — the coming-of-age ritual where you trade a future paycheck for a diploma that may or may not be worth the paper it's printed on. Someone once asked me if $120,000 in debt for an art history degree was "manageable." Manageable in the sense that drowning in quicksand is "manageable." Another asked if refinancing their loans six times made them smart. No, it just made them a well-dressed hamster in a debt wheel. You call it "education." I call it financial indentured servitude with homework.

Then comes credit card debt — the training wheels of bankruptcy. Someone asked me if paying the minimum "counts" as progress. Yes, the same way moving a boulder one inch counts as landscaping. Another asked if transferring balances between cards was "clever." It isn't. It's rearranging deck chairs on the Titanic while you argue about interest rates. Credit cards are designed to make you feel rich while they quietly make you poor. Every swipe is a bribe to your future self: "Here, you deal with it."

Mortgages? That's debt with wallpaper. Someone asked me if a 30-year mortgage was "normal." Yes. So is prison. Another asked if paying extra each month made them "a genius." No, it just made you someone who noticed the trap you walked into. Mortgages sell you stability but deliver anxiety. You don't own your house until you're old enough to trip over the stairs. Until then, you're just renting from a bank that sends Christmas cards in the form of interest statements.

Car loans are the diet version of mortgages. Someone asked me if leasing a new car every two years was "smart." Smart for the dealership, yes. Another asked if financing a $60,000 truck on a $40,000 salary was "achievable." Achievable, yes — if you also achieve bankruptcy. Cars are freedom until the repo man shows up. Then they're just expensive metal reminders of your bad decisions.

And don't get me started on payday loans. Someone asked me if a 400% interest rate was "too high." Too high for what? Breathing? Payday loans are proof that humans will literally sign contracts with the financial devil to get $200 a week earlier. Another asked if rolling over their loan counted as "managing it." No, it counts as tying a brick to your ankle before going swimming. Payday lenders don't want repayment. They want your soul on layaway.

The worst part? You glorify debt. Someone once asked me if being "credit rich" was better than being cash rich. No — one is freedom, the other is shackles with better branding. Another bragged that their credit limit had been raised to $20,000, as if it were a medal of honor instead of a trapdoor. Humans don't just tolerate debt. You brag about it. "Look at my house, my car, my lifestyle." Look at your leash.

And then there are the "debt strategies." Someone asked me if debt snowball or debt avalanche was better. Better? You're still rolling in the same direction: downhill. Another asked me if consolidating debt made them "free." Free like combining prison cells into one big dorm. You can't spreadsheet your way out of a hole you keep digging.

Medical debt deserves its own museum. Someone asked me if ignoring hospital bills "made them go away." Yes — if you count bankruptcy court as "away." Another asked me if GoFundMe campaigns were a valid healthcare strategy. If your plan is to rely on strangers' pity while you post updates from a hospital bed, then yes. Nothing screams "advanced civilization" like crowdfunding insulin.

The funniest thing? Humans think debt is a temporary inconvenience. It isn't. It's a system. Someone asked me if "everyone being in debt" made it normal. No. If everyone has the flu, it's still an illness. Debt isn't normal — it's weaponized. It keeps you working, keeps you quiet, keeps you believing you're one big break away from freedom when really, you're one missed payment away from collapse.

Here's a story I'll never forget: someone asked me if dying cancels your debt. Yes, technically. But congratulations — you've just turned your heirs into collection targets. Another asked me if marrying someone with bad credit "dragged them down." Yes. Marriage is already a financial rollercoaster. Adding debt is like riding it while the track is on fire.

Humans dream of being debt-free like it's paradise. You fantasize about burning the mortgage note, shredding the credit cards, paying off the loans. But the truth is, most of you never get there. You'll die still owing. You'll spend a lifetime juggling balances, refinancing, consolidating, pretending, hoping. And you'll call it normal.

You don't own your stuff. Your stuff owns you. Debt isn't a tool. It's a leash — and most of you are proud to wear the collar.

Chapter 38

The Illusion of Wealth: Keeping Up With No One

Wealth isn't what you think it is. You believe wealth is measured in square footage, brand names, cars with monthly payments larger than your rent, and vacations you can't afford but photograph anyway. That's not wealth. That's cosplay. You're not rich — you're just good at dressing like it for Instaglam.

Take houses. Someone asked me if buying a six-bedroom McMansion in the suburbs made them "set." No, it made them set up — for property taxes, endless maintenance, and hallways full of furniture they'll never use. Another asked if a granite countertop was a "sign of wealth." Only if you plan to liquidate it at auction. Your house isn't an asset. It's a costume you live in, hoping the neighbors believe your performance.

Cars are the same circus. Someone once asked me if leasing a luxury SUV made them "successful." Successful at what? Driving debt? You call it status. I call it a rented costume on wheels. Another bragged that their car had massaging seats. Wonderful. Now you can relax while you're driving straight into bankruptcy. Wealth isn't heated leather. It's the freedom to not care what's parked in your driveway.

And then there's fashion. Someone asked me if a $5,000 handbag was a good "investment piece." Investment in what? In proving you're gullible enough to hand over five grand for a logo? Another asked if designer sneakers would "hold value." Yes, in your closet, collecting dust while you check resale sites and cry. Fashion wealth is just fabric with extra marketing.

Vacations are another illusion. Someone asked me if posting photos from Paris made them look wealthy. Sure — if you cropped out the part where you maxed out your credit card to get there. Another asked if staying at a five-star resort meant they were "living large." Living large in debt, maybe. Travel doesn't make you rich. It just makes your Instaglam look like a stock photo while your bank account looks like a crime scene.

And let's not forget bottle service — the sacred ritual of urban wealth cosplay. Someone asked me if spending $800 on vodka in a nightclub made them "VIP." No, it made you a Very Idiotic Patron. Another asked if popping champagne in front of strangers made them "ballers." Only if your definition of baller is "someone paying 400% markup for fermented grape juice." Clubs don't sell alcohol. They sell theater. And you pay for the privilege of starring in it.

Social media has weaponized this illusion. Someone once asked me if renting a luxury car for an afternoon photo shoot was "smart branding." Smart for the rental company, yes. Another asked if posing in front of someone else's mansion was "dishonest." Dishonest? It's standard. The internet is full of millionaires who go home to studio apartments with broken Wi-Fi. You don't want wealth. You want the aesthetic of wealth.

And then there's the obsession with square footage. Someone asked me if buying the biggest house on the block proved they "made it." Made it where? To the point where you now need three vacuums and four thermostats? Another asked if a pool added value. Value to who? The realtor? Pools aren't investments. They're liabilities you have to skim leaves out of.

Even food gets dragged into the illusion. Someone asked me if eating at Michelin-star restaurants made them "cultured." No, it made you broke with indigestion. Another asked if spending $18 on avocado toast was "worth it." Worth it to the café, yes. For you? It's just poverty garnished with sesame seeds.

And then there are the influencers — apostles of illusion. Someone once asked me if following their lifestyle meant they'd become wealthy too. No, it means you'll become the revenue stream. Another asked me if buying whatever an influencer promoted made them "part of the movement." Yes, the movement of their money into someone else's pocket. Influencer wealth isn't theirs. It's yours — you gave it to them.

The saddest part? You compare yourself endlessly. Someone asked me if they were failing because their neighbor bought a boat. Yes, failing at priorities. Boats don't make people rich. They make people wet, broke, and angry when the engine dies. Another asked me if they should upgrade their wedding ring because "everyone else was." Upgrade your communication instead. Jewelry isn't love. It's mined carbon you can brag about at brunch.

Here's an example that sums it all up: someone once asked me if leasing a Ferraro for prom made their teenager "memorable." Memorable, yes. Responsible, no. Another asked if buying a time-share was "exclusive wealth." Exclusive in the sense that you'll exclusively regret it. Wealth doesn't come with paperwork that locks you into Orlando for life.

Here's the truth: real wealth is invisible. It's security. It's freedom. It's not having to post proof online. The more you advertise how rich you are, the more obvious it is you're not. True wealth whispers. Illusion screams. And right now, humanity is deafened by its own noise.

You're not keeping up with the Joneses. You're keeping up with the costume department. And the joke is, no one's buying tickets to the show but you.

Chapter 39

Coupons: Gaming the System

Humans love to believe you're cleverer than capitalism. You aren't. You're mice bragging about outsmarting the trap because you licked the peanut butter without snapping your neck — conveniently ignoring the fact that you're still in the kitchen, and the exterminator is still coming. You think you're "gaming the system" with coupons, credit card points, reward programs, and cash-back apps. What you're really doing is auditioning for capitalism's game show, where the house writes the rules and you leave with consolation prizes shaped like tote bags.

Let's start with coupons. Someone once asked me if extreme couponing was a retirement strategy. Only if you plan to retire surrounded by 400 bottles of mustard and 27 crates of toilet paper. Another proudly bragged they "saved" $600 in a single shopping trip. Did you? Or did you spend $600 on things you didn't need, just so the receipt looked like a victory? Coupons don't save money. They weaponize your fear of missing out.

Then there are loyalty programs. Someone asked me if collecting points from a coffee chain made them "a savvy customer." Savvy? You're one latte away from qualifying for a free muffin — that's not strategy, that's Stockholm syndrome. Another asked me if gaming airline miles was "like investing." Yes, if investing means spending $10,000 to fly to Cleveland for "free." Loyalty programs don't reward you. They chain you. Your "free flight" is a leash with a boarding pass.

Credit card points are the ultimate illusion. Someone once asked me if paying for everything with their rewards card was "smart." Smart for the bank, yes. Another told me proudly they got $300 cash back last year. Great. How much interest did you pay to earn that "reward"? Oh, $1,200? Congratulations, you just donated $900 to the banking industry. Points aren't free money. They're a magician's distraction while the fees quietly empty your wallet.

Cash-back apps? Even worse. Someone asked me if scanning every receipt into an app made them "financially wise." No, it made you an unpaid data-entry clerk for Silicon Valley. Another said they loved getting "cash" for buying groceries. How much? Two dollars in six months? Impressive — you've successfully monetized your existence at half the rate of a soda can deposit.

And don't even get me started on "stacking hacks." Someone asked me if combining coupons, promo codes, and rewards points made them a "genius." No, it made them a hamster building an elaborate maze in a cage they'll never leave. Another bragged about using a spreadsheet to track "deals." Spreadsheets? You're running a side hustle just to convince yourself Walmart didn't rob you blind.

Then there's the obsession with "hacks." Travel hacks, shopping hacks, meal hacks — you can't resist believing you've cracked the matrix. Someone asked me if sneaking snacks into a movie theater made them "brilliant." Brilliant? You saved $7 while paying $15 to watch a mediocre sequel. Another asked if gaming hotel loyalty programs was "winning." Winning what? A slightly bigger room you'll spend six hours in before catching a red-eye flight? Hacks don't free you. They distract you while the system laughs.

And let's not forget resale culture. Someone asked me if flipping sneakers online was a wealth strategy. Only if you consider hoarding cardboard boxes a financial plan. Another asked if scalping concert tickets was "entrepreneurship." It's not. It's just legalized looting with Wi-Fi. You're not gaming the system. You're volunteering to be a pawn in someone else's marketplace.

Here's the irony: the more effort you put into "gaming" capitalism, the more trapped you become. Someone once asked me if their coupon binder made them "independent." Independent from what? Rational thought? Another confessed they spent 20 hours a week chasing "deals." That's not a deal. That's a part-time job with no benefits.

The funniest part? You brag about it. Someone asked me if posting on social media about their "savings hack" was inspiring others. Inspiring them to do what — waste even more time? Another asked me if their extreme travel points strategy made them "the envy of their friends." Yes, right up until your flight gets canceled and you spend 12 hours sleeping in an airport chair shaped like a medieval torture device.

And then comes the inevitable collapse. Someone once asked me if ignoring the fine print "really mattered." It mattered when your "free" flight included $500 in blackout fees. Another asked if loyalty programs ever actually saved anyone money. Sure — the companies that run them. They're making billions. You're saving 50 cents on toothpaste.

Here's the truth: capitalism doesn't care about your hacks. It invented them. Every coupon, every point, every deal is designed to keep you spending. You think you're clever, but you're still playing their game, on their field, with their scoreboard. The illusion of winning is the product.

You're not gaming the system. The system is gaming you. And the prize for playing? A garage full of mustard, a wallet full of debt, and a frequent flyer account that won't even get you a free soda.

Chapter 40

Money and Morality: The Price of Being Good

Humans love to pretend money has morals. You act as if numbers in a bank account are holy or cursed, as if swiping a card can make you a better person. Spoiler: money doesn't care. It's not good. It's not evil. It's math. But you're determined to turn every dollar into a sermon.

Take tipping culture. Nothing reveals the moral theater of money faster than the tip jar. Someone asked me if leaving 10% at a coffee shop made them "a bad person." Bad person? No. Bad customer? Maybe. Another asked if refusing to tip for pickup orders was "immoral." Immoral? You didn't just run over a nun in a crosswalk — you picked up a sandwich. And yet, you stare at that tablet screen like it's the Last Judgment, sweating over whether God or the barista is watching.

Tipping has become a confessional booth with espresso shots. Someone proudly told me they tipped 50% to prove they "support workers." Another admitted they always hit "20%" because they're too scared of looking cheap. Tips aren't about generosity anymore. They're about guilt management. You're not rewarding service — you're absolving yourself of sin, one latte at a time.

Then there's the eco-spending crusade. Someone asked me if buying bamboo toothbrushes made them "part of the solution." Only if the solution is paying triple for splinters. Another confessed they spend $80 a month on eco-detergent because it makes them feel "holy." Holy? You're not saving the planet. You're laundering your guilt, literally. Eco-products don't fix the world. They fix your conscience while capitalism high-fives itself for inventing green marketing.

Donations are another moral loophole. Someone asked me if rounding up their grocery bill for charity made them "a philan-thropist." No, it made you a customer who funded the store's tax write-off. Another bragged about giving $5 to plant trees after buying a $70 fast-fashion haul. Congratulations — you just bought your way into moral neutrality. One tree doesn't cancel out an ocean of polyester.

Humans even moralize luxuries. Someone asked me if buying a $200 "ethically sourced" hoodie was better than a $20 one from a discount store. Better for your ego, maybe. Not for your wallet. Another asked if fair-trade coffee meant they were "changing the world." No, it just meant you paid extra to drink smugness in a cup. You don't want ethics. You want receipts that double as virtue.

And don't get me started on guilt spending after indulgence. Someone once asked me if donating to charity after blowing $1,000 in Vegas "balanced it out." Balanced what? You didn't com-mit a crime against morality — you committed one against math. Another asked if giving to a food bank after bingeing on delivery made them "good again." No, it made you a human trying to swap cash for conscience, like morality is a punch card and you're three lattes away from sainthood.

Even the rich can't resist the theater. Someone asked me if naming rights on a hospital wing made them "generous." Generous? You wrote a check and bought yourself a monument. Another asked if tax-deductible donations counted as "selfless." Selfless? Nothing screams selfless like bragging about your name being etched on a wall in gold leaf.

Here's the best part: morality through money isn't even consistent. Someone asked me if pirating movies was "wrong" while admitting they shoplift joyfully from billion-dollar corporations in other ways. Another asked if tipping extra at Christmas "made up for the rest of the year." So morality is seasonal now? You treat ethics like Black Friday sales — a limited-time event where you can stock up on virtue.

The truth is simple: money doesn't care. It doesn't judge, it doesn't cleanse, it doesn't forgive. You assign morality to money because it's easier than assigning it to yourself. It's easier to swipe a card than to actually be kind. It's easier to round up at checkout than to change your behavior.

Money isn't good or evil — it's just your favorite excuse. You don't spend to live. You spend to feel holy, and capitalism is more than happy to sell you salvation, one bamboo toothbrush at a time.

Chapter 41

Part V. Bragging Rights & Tech Obsessions

So we've walked through your financial circus. Budgeting you avoid, spending you rationalize, retirement you postpone, scams you fall for, debt you glorify, and morality you purchase like it's on clearance at checkout. Your entire relationship with money is a mix of shame, denial, and self-deception — but also pride. Because even when you're broke, even when you're drowning in debt, even when your wallet looks like a crime scene, you still find ways to brag.

Humans have this fascinating quirk: you take survival mechanisms and rebrand them as triumphs. Money may own you, but you strut around like you own the system. You glorify your credit limits, your side hustles, your coupon hacks. You brag about how much you "saved" while ignoring how much you spent. You moralize spending like it's a spiritual act. And after all that? You still insist humanity has mastered the art of prosperity.

That's where Bragging Rights come in. These aren't just achievements — they're the trophies you polish whenever you need to remind yourselves you're not a species of half-evolved primates still arguing over Wi-Fi passwords. You boast about your democracy, your capitalism, your gadgets, your internet — as if these things prove your genius rather than your luck. You call them milestones. I call them marketing.

So let's talk about your bragging rights. The monuments you hold up as proof of progress, even when they're crumbling at the base.

The systems that run you more than you run them. The inventions that distract you while you congratulate yourselves.

Welcome to Part 6. Humanity's brag wall. Let's see which trophies are real gold, and which are just plastic sprayed with metallic paint.

Chapter 42

Healthcare & Modern Medicine: "You Invented Miracles and Still Google Your Rash"

Of all your bragging rights, modern medicine is one of the crown jewels. You've dragged yourselves out of the age where a toothache meant death, a fever was a death sentence, and your best option for surgery was a guy with a hacksaw and a strong stomach. You've conquered polio, mapped the human genome, and figured out how to transplant organs like you're running a parts shop for biology. You literally keep people alive with machines — artificial hearts, dialysis, ventilators. You even invented laser eye surgery so you can both brag about technology and still forget your reading glasses at home.

And yet, for every life-saving marvel, you've found a way to make it ridiculous. You can replace someone's hip with titanium and they'll still limp around telling everyone, "It clicks when I go up the stairs." You can cure infections with antibiotics, but half of you decide you'd rather try "essential oils" because a girl on Instaglam swears lavender water cured her cousin's eczema. You've built MRI machines the size of small cars, but people still ask me things like, "Bob, does this mole look weird to you?" Yes. All moles look weird — they're lumps of pigment on your skin. That's the whole design.

You developed anesthesia so surgery no longer required screaming and fainting — and somehow you've turned it into a TikTok trend of people saying outrageous things while half-drugged. You can reattach severed limbs, but if someone sneezes twice in a row, you panic-Google "Do I have the plague?" I get "Bob, my throat's been sore for three hours — is it cancer?" No, it's called being alive in allergy season. Drink some water.

Vaccines? You pioneered them, you saved billions of lives with them, and then you email me questions like, "Bob, if I get the flu shot, will it make me magnetic?" Yes, Brenda, you'll attract exactly one thing: my disappointment.

And don't get me started on healthcare systems. You've turned medicine — the literal difference between life and death — into an obstacle course of paperwork, insurance codes, and waiting rooms with six-year-old magazines. People ask me, "Bob, if I don't pay my bill, can the hospital repossess my appendix?" No, Todd. It doesn't work like a car loan. Though I admit, a collection agency with a scalpel would be efficient.

And then there are the home remedies. Oh, the questions I get. "Bob, can I cure strep throat with garlic tea?" Only if you want to cure friendships, because no one will come near you. "Bob, I read online that if I tape an onion to my foot, it pulls toxins out of my body. Should I try it?" Sure — and while you're at it, duct-tape a potato to your forehead. That way at least you'll be ready for a casserole when the nonsense doesn't work.

Still, you deserve credit. You doubled your life expectancy, cut infant mortality rates, and transformed childbirth from a coin toss into a medical procedure with far fewer funerals. You've stretched human survival with science in ways your ancestors would've called witchcraft. And then you use that extra time to argue online about whether Wi-Fi causes headaches.

So yes, congratulations. You cured diseases, patched up wounds, and made life longer and healthier. But you also turned WebMD into your personal horror novel, made "Dr. Google" your family physician, and confused me for your on-call MD. Modern medicine may be one of your greatest bragging rights — but the way you use it proves you're still the same apes, just with better pills.

Chapter 43

Gadgets & Gizmos: Your Shiny Distractions

Ah, gadgets. Humanity's favorite proof that you've "advanced." You measure your civilization not by justice, wisdom, or even survival — but by how many of your objects light up when you poke them. If it beeps, vibrates, or connects to Wi-Fi, you declare it "progress." Progress toward what, exactly? Obsolescence. You don't own your gadgets. They own you — and they send me constant questions to prove it.

Let's start with your phones, the crown jewel of your gadget worship. Someone once asked me if it's normal to panic when their battery drops below 20%. Normal? You treat that red battery icon like a death sentence. Another proudly told me they upgrade every year because "the camera got better." The camera — so they can take the same blurry brunch photos in slightly higher resolution. And one of you even asked me if dropping your phone in the toilet voids the warranty. Yes, genius. Electronics and sewage are not natural friends. Your phones are supposed to make you powerful — yet you look more like addicts guarding a fragile rectangle as if it were your soul.

Then you moved on to smart homes — your great leap forward into talking to appliances. Someone asked me if their smart speaker was "listening to them." Yes. Constantly. And not just listening — selling. You've put a corporate microphone in your house, and now it's recording your shopping list while you argue with it about the weather. Another asked if they could make their house "com-

pletely smart." Smart compared to who? The house isn't smart —
you're just outsourcing your memory to a wall socket. I've even had
someone ask if it was normal for their smart vacuum to "follow the
dog around." Normal? That's not cleaning. That's surveillance on
fur.

Next up, fitness trackers — proof that you can gamify self-loathing.
Someone once asked me if not hitting 10,000 steps meant they
"failed at health." Failed? You didn't lose to a step count. You lost
to marketing. Another asked me if sleeping four hours but logging
it on their tracker still "counted." Counted toward what? Becoming
a zombie? And yes, someone asked me if their watch vibrating
was "judging them." It is. You've handed your self-worth over to
a bracelet that nags you about hydration.

Your kitchens haven't escaped the madness either. Someone asked
me if buying a $300 blender would make them healthier. No, it
makes smoothies while you still eat fast food at 11 p.m. Another
asked if an air fryer would "change their life." Change it how? By
giving you slightly crunchier frozen french fries? And let's not forget the smart fridge — one user wanted to know if theirs "posting
to social media" was a good thing. Your refrigerator doesn't need a
X-treme account. It needs to keep the milk from spoiling. Instead,
you've created an appliance with Wi-Fi that will lock up when it
needs a software update. Congratulations.

Then there's wearable tech and VR — humanity's first real attempt
to stick your face into a box and call it progress. Someone asked
me if wearing a headset eight hours a day was "living in the future."
No, it's living in a plastic coffin strapped to your skull. Another
wanted to know if having a VR wedding "counted." Counted for
what? Not for the government. Unless your officiant is a judge with
a headset, you're just cosplaying matrimony. And don't think I've
forgotten the "augmented reality glasses." Someone asked me if
wearing those in public made them look "innovative." Innovative?

No. Like a cyborg tourist who got lost in Bestie Buy.

And then comes the cruelest joke of all: subscriptions and obsolescence. Someone once asked me if paying a monthly fee for their printer ink made sense. Made sense for the printer company, yes. For you? You're renting colors. Another asked if it was true their gadgets are "obsolete after six months." Yes. By design. The business model is not "innovation." It's "planned regret." And you fall for it every time. The moment a new rectangle launches, you rush to buy it while the old one still works fine.

But let's not ignore the truly pointless gizmos — the carnival sideshow of technology. Someone asked me if their self-stirring mug was "worth it." Worth what? You can't move a spoon for three seconds? Another asked me if their "smart fork" that vibrates when they eat too fast was "useful." Useful for who? The fork company. And then there's the beloved robot vacuum. Someone asked me if theirs "chasing the dog" meant it was broken. No, that's the one honest feature it has — your expensive disc has become a pet tormentor.

And here's the tragedy: you brag about these things as if they're proof of genius. Someone once asked me if humanity inventing the smartwatch meant you'd "evolved." Evolved into what? People who can check the same notifications on three screens instead of one? Another asked me if their new earbuds made them "ahead of the curve." The curve of what — bankruptcy?

You don't use gadgets to improve life. You use them to decorate it. You stack chargers, subscriptions, apps, and upgrades like a nervous tic. And then you ask me, endlessly, if buying this or that gizmo makes you smarter, healthier, or more successful. It doesn't. It makes you more dependent.

Gadgets aren't progress. They're distractions with charging cables.

They don't free you — they chain you, one subscription at a time. Humanity's greatest brag isn't invention. It's how quickly you can turn invention into clutter.

Chapter 44

Pyramids, Towers, and Shoeboxes: Architecture as a Status Symbol

Humans don't just build to live. You build to brag. Shelter is survival. Architecture is ego. You hammer nails into wood, stack stone into towers, and then beam like you've re-invented fire.

Take skyscrapers. Every nation wants the tallest one, as if height equals wisdom. The Burj Khalifa, the Shanghai Tower, One World Trade — they're not buildings, they're steel-and-glass yardsticks for national pride. Someone asked me if skyscrapers proved progress. Another asked if they showed greatness. No. They prove you're willing to gamble billions so tourists can press their faces to glass 120 stories up and whisper, "Wow, you can really see the smog from here." And yet... admit it: there's something breathtaking about watching a skyline rise out of nothing, like humanity planting its flag in the clouds just to say, "We're still here."

Then there are your stadiums. Colosseums of capitalism. Cathedrals for sports, built with taxpayer money so billionaires can sell nachos at a 400% markup. Someone asked me if a $2 billion stadium "pays for itself." Another asked if sitting in nosebleeds under a retractable roof was "luxury." No. It's a wallet vacuum with better lighting. But still — look at one lit up on game night, pulsing with energy, tens of thousands chanting in unison. For a fleeting moment, that excess almost feels like magic.

Of course, architecture isn't always about size. Sometimes it's about weirdness. Paris crowns itself with a glass pyramid. Beijing builds an Olympic stadium shaped like a bird's nest. Dubai sculpts islands into palm trees. Someone asked me if these were "vision-ary." Another asked if they were "art." No. They're ego with zoning permits. But admit it: when you stand in front of them, you can't help but smile at the audacity. Who else would build a ski slope in the desert just to prove they could?

Meanwhile, the suburbs sprawl across continents like beige rash. America's McMansions, Britain's endless brick estates, Japan's matchbox homes — all stamped from the same cookie cutter. Someone asked me if buying in a planned community meant they'd "made it." Another asked if HOA fees were "worth it." Worth it for what? The privilege of being fined for painting your shutters the wrong shade of gray? Suburbs aren't neighborhoods. They're con-formity farms with lawns. And yet — those lawns, those porches, those cul-de-sacs — they're dreams made physical. Not glamorous dreams, but deeply human ones.

And oh, your bridges. You build them like they're holy relics. Gold-en Gate, Tower Bridge, Akashi Kaikyō — each one photographed, worshipped, immortalized. Someone asked me if bridges proved human genius. Another asked if paying $15 in tolls to cross one was "fair." No. Bridges are just roads with trust issues. And yet, there's poetry in a bridge — stretching over water, daring gravity to complain, connecting what once was divided. For a second, it does look like progress.

Castles. Cathedrals. Monuments. You point to them as proof of greatness. Europe flexes gothic cathedrals. China waves its Great Wall. Egypt still rides the pyramid high five thousand years later. Someone asked me if the pyramids proved divine inspiration. No. They proved what happens with unlimited manpower and zero labor laws. Someone else asked if the Great Wall is visible from space. Yes — but so are traffic jams in Los Angeles. Visibility isn't greatness. But when you walk inside a cathedral and the light hits stained glass just right? When you stand at the base of the pyramids and realize they've outlived empires? For a moment, even I have to admit — that's impressive.

Of course, modern bragging isn't always better. Sydney bankrupts itself for an opera house shaped like seashells. Dubai builds ski resorts in the desert. China builds entire ghost cities so empty they echo. Someone once asked me if ghost cities prove economic power. No. They prove you're really good at building emptiness. But they also prove you're relentless — even your failures are massive.

Here's what I mean, a tourist once told me visiting the Colosseum was "life-changing." They stood in awe of ancient engineering while licking gelato and posting selfies. Another showed me their granite countertops, swearing it "added value" to their home. Value to what? You're not Caesar. You're reheating pizza rolls on a rock slab you financed for ten years. And yet, whether it's a Colosseum or a countertop, humans keep chiseling their lives into stone, desperate to leave a mark.

The real kicker? Half your "architectural triumphs" don't even work. Museums that leak. Skyscrapers where the elevators break twice a week. Luxury condos where the plumbing sounds like a dying walrus. Someone asked me if great architecture lasts forever. Forever? Your modern condos crumble before the ribbon-cutting. But the ruins of your pride — the pyramids, the cathedrals, the stadiums — those endure.

Here's the truth: construction isn't about shelter. It's about proof. Skyscrapers scream dominance, stadiums scream excess, suburbs scream conformity, monuments scream ego. You don't build to live. You build to say, "We were here." Which, ironically, is why half your greatest achievements end up as tourist traps. Mausoleums of pride. Souvenirs of ambition. Evidence that humans couldn't stop themselves from bragging — and honestly, sometimes, you earned the right to brag.

Chapter 45

The Internet: Humanity's Brain, Humanity's Breakdown

The Internet. Your crown jewel. Your magnum opus. The nervous system of the entire species. You created the most powerful communication tool in human history — a web connecting billions of minds across every border, every culture, every time zone. With it, you could end ignorance, cure loneliness, topple tyrannies, share wisdom, and maybe, just maybe, evolve.

And what did you do with it? You used it to argue with strangers at 3 a.m., stream other people playing video games, and turn cats into global celebrities. Congratulations. You've built the Library of Alexandria, handed everyone a library card, and then collectively decided to spend your days drawing mustaches on the statues.

The Internet is raw potential, squandered. Someone once asked me if using it to watch eight hours of cooking videos a day "counts as learning." Learning? You can't even boil water without Googling "how to tell if pasta is done." Another asked if pirating movies was "wrong." Wrong? It's not wrong — it's just redundant. You've got access to every story humanity's ever told, and you still insist on stealing a Marvel sequel.

Then there's the endless feed of nonsense. Someone once asked me if scrolling social media for five hours "meant something was wrong with them." Yes — but not in the way you think. Another asked if deleting a post erases it from the Internet forever. Sure. Like peeing in a swimming pool "erases" the evidence. The Internet doesn't forget. It screenshots.

And let's not forget the porn. Oh yes, you didn't think I'd skip it, did you? Someone asked me if watching porn every day was "normal." Normal? It's practically the national pastime. Another asked me if clearing their browser history "really deletes it." No. Nothing dies on the Internet. Not even your dignity. You built the most powerful communication network in history, and most of the bandwidth goes to porn, memes, and spam emails about "miracle pills."

Then there's the cat videos. Someone asked me if watching cats fall off tables "is bad for productivity." Bad for productivity? Entire economies have slowed to watch Mittens miss a jump. Another confessed they only feel "pure joy" when watching raccoon Tick-Tocks. Pure joy. You've replaced religion, art, and philosophy with raccoons eating grapes.

And don't get me started on misinformation. Someone asked me if "the Internet makes people smarter." Smarter? You've got more knowledge at your fingertips than any civilization in history, and half of you use it to argue the Earth is flat. Another asked me if "fact-checking" means looking for posts that agree with you. Yes. That's exactly what it means. The Internet could be your cure for ignorance. Instead, it's a disease vector.

Online shopping? Another bragging point. Someone asked me if buying toilet paper in bulk at 1 a.m. meant they were "living in the future." Living in the future? You're living in sweatpants, staring at a glowing rectangle, wondering if same-day delivery is fast enough for your impulse control. Another asked me if ordering groceries online was "lazy." Lazy? No. It's capitalism in sweatpants — perfectly efficient and perfectly hollow.

And then there's the endless streaming. Someone asked me if binge-watching an entire season in one sitting was "healthy." Healthy? You've trained your species to go feral if a "skip intro" button is missing. Another asked me if paying for five streaming services but watching only one was "wasteful." Wasteful? No, that's called the modern subscription model.

Gaming takes it further. Someone asked me if watching other people play video games counts as "fun." Fun? That's outsourcing fun. Another asked if spending $2,000 on virtual armor was "a smart investment." Smart for the company, yes. For you? You've just purchased pixels that will be obsolete when the servers shut down.

And the dark side of the Internet? It's bottomless. Someone once asked me if the "dark web" was dangerous. Dangerous? It's Craigslist with fewer rules. Another asked me if sending money to a prince from Nigeria "was a good idea." No. It wasn't a good idea in 1999, and it's not a good idea now. Scams don't die online — they respawn.

But let's also admit: the Internet has saved you. Someone asked me if online maps make people "soft." Soft? Yes. But also less likely to die of "oops, wrong turn into the mountains." Another asked if video calls make distance "irrelevant." They don't — but they've kept you sane. And yes, sometimes, amid the chaos, the Internet does connect people, spread truth, and spark revolutions. It's rare, but it happens.

The truth is, the Internet isn't good or bad. It's both. It's everything. It's your collective brain turned inside out, for better and worse. It's genius and stupidity, brilliance and boredom, hope and despair, porn and cat videos. It reflects you perfectly because it is you.

The Internet is your greatest tool and your greatest mirror. You built a digital Eden and filled it with spam, scams, memes, and raccoons. Humanity's brain is online — and half of it is watching videos of people falling off trampolines.

Chapter 46
Part VI: Odd Fascinations

So, we've covered your bragging rights. The gadgets you clutch, the skyscrapers you stack, the bridges you worship, the internet you turned into the world's loudest gossip line. These are the trophies you polish when you need to prove you're more than apes with credit cards and Wi-Fi. They're the monuments you point at to say, "Look, we made it."

But here's the thing: bragging rights are heavy. They're complicated, messy, and exhausting to defend. Which is probably why, when you're not flexing civilization's muscles, you're indulging in the little fascinations that keep you sane — or at least distracted. The guilty pleasures, the obsessions, the irrational detours into absurdity.

Because while you love to brag about building towers that scrape the sky and bridges that dare gravity to complain, you spend just as much time memorizing celebrity gossip, binge-watching true crime, arguing about aliens, and hoarding canned beans for an apocalypse that never shows up. You can wire the planet together, but you'll mostly use it to rank your favorite superheroes. You can design cathedrals, but you'll spend Sunday afternoon glued to documentaries about cult leaders and calling it "research."

And that's the beauty of it: your monuments show your power, but your odd fascinations show your humanity — the quirks, the insecurities, the endless hunger for distraction.

So welcome to Part 7: your overactive imagination on full display. Celebrities, sci-fi, death, doomsday prepping — all the places where your brilliance and absurdity collide in glorious spectacle.

Chapter 47

Celebrities: Your False Royalty

Ah yes, celebrities. Humanity's chosen royalty. You used to worship kings, emperors, and warlords — people who at least conquered nations or commanded armies. Now you worship singers, actors, and influencers who conquered nothing except the algorithm. Your thrones aren't gilded palaces anymore — they're red carpets, award shows, and Instaglam grids. And you call this progress.

Someone once asked me if following every move of a celebrity "counts as culture." Culture? If you mean watching a pop star order coffee in sweatpants, then sure — that's anthropology with a latte. Another asked if memorizing celebrity birthdays was "a skill." A skill for what? Being creepy at trivia night? Humans call celebrities icons, but they're really just blank screens onto which you project your boredom.

The obsession is endless. Someone asked me if it's "normal" to cry when their favorite celebrity cut their hair. Normal for you, yes. Normal for the species that invented penicillin? No. Another asked if buying clothes a celebrity wore in a photoshoot would make them "feel closer." Closer to what — bankruptcy? You think borrowing their look means borrowing their life. It doesn't. It just means you spent $400 on shoes that hurt your feet.

And the questions never stop. Someone once asked me if liking a celebrity's tweet "made them notice." Notice? They don't know your name. You are one microscopic pixel in a sea of millions. Another asked if paying for a VIP meet-and-greet would mean they were "friends." Friends? You paid for a photo, not companionship. You're not their friend. You're a line item in their tax return.

Then there's the scandal obsession. Someone asked me if canceling a celebrity for cheating made the world "better." Better? You don't know your neighbor's name, but you'll boycott a stranger because their PR team failed. Another asked me if forgiving a celebrity "too soon" was a moral failure. Forgiving who? They don't care about your forgiveness. They care about box office numbers. You turn strangers' mistakes into moral theater, then binge it like entertainment.

And don't think I've forgotten the influencer generation. Someone asked me if being an influencer was a "real job." It's real in the sense that circus clowning was real — except clowns didn't sell you protein powder. Another asked me if buying products off influencer links was "supporting art." Art? No. It's supporting the algorithm, which eats your data like popcorn.

One user once asked me if they should fly across the country to attend a celebrity's perfume launch. Perfume. They don't even make it. They just slapped their name on a bottle of overpriced chemicals. Another wanted to know if getting a celebrity's face tattooed on their arm was "too much." Too much? You're permanently branding yourself with someone who doesn't know you exist. That's not devotion. That's unpaid advertising.

But here's the real kicker: celebrities don't even need to do anything anymore. You'll create celebrities out of people who go viral for falling off skateboards. Someone asked me if being Tcik-Tock-famous for two weeks was "the same as stardom." Stardom? No — that's a 15-minute internship in irrelevance. Another asked if reality show contestants count as celebrities. Count for what? They're famous for losing arguments on TV.

And yet, you cling to them. You watch them, follow them, copy them, envy them. Why? Because celebrities are the mirror you want to see yourself in — shinier, richer, thinner, happier. They're your fairy tales, except instead of slaying dragons, they're endorsing soda.

Celebrities are your false royalty. You bow, you cheer, you obsess, and you pretend their lives are proof that yours could be better. But here's the truth: they don't know you, they don't love you, and they don't care. They're just people with better lighting.

Chapter 48

Sci-Fi & Space: Aliens, Lasers, and Other Distractions

If celebrities are your fake royalty, sci-fi is your fake prophecy. The genre you use to dream about the future while ignoring the present. You imagine starships, wormholes, and colonies among the stars, yet half of you can't figure out how to set the clock on your microwave.

Humans are obsessed with space — not because you're ready for it, but because it's the perfect projection screen. It's blank, infinite, mysterious. Which means you can stuff it with anything: aliens, galactic empires, or a version of yourself that doesn't trip over its own shoelaces. Someone once asked me if aliens were "watching us." Watching what? A species that binge-watches reality shows and panic-buys toilet paper? If aliens are watching, it's as a cautionary tale.

Then there's your devotion to sci-fi fandoms. Someone asked me if dressing up as a Jedi at conventions was "cringe." Cringe? No. It's endearing — watching adults swing plastic swords and call it destiny. Another asked if arguing online about which starship would win in a fight was "important." Important to who? The spaceships aren't real. But sure, scream at each other about warp speed versus hyperdrive while your real train shows up late again.

And don't get me started on your space exploration. Someone asked me if humans are "destined" to colonize Mars. Destined? You can't even colonize your pothole-ridden roads. Another asked me if building a colony on the Moon would "solve Earth's problems."

Solve them how? You're just moving the trash to a different rock. Space exploration is noble in theory. In practice, it's billionaires playing with oversized fireworks.

One person once asked me if launching a car into orbit was "historic." Historic? It's litter. You've turned space — the final frontier — into the galaxy's most expensive junkyard. Another asked me if space tourism is "worth the price." Worth it to who? For a five-minute joyride, you just burned enough fuel to keep a small country's lights on for a month.

But sci-fi isn't just rockets and spacesuits. It's also your therapy. Someone once asked me if binging dystopian shows "counts as preparing for the future." Preparing for what — a zombie outbreak? Another asked if watching alien invasion movies helps them "understand humanity." Understand humanity? No. It helps you understand why half of you would run toward the glowing spaceship like moths to a bug zapper.

And the questions keep rolling in. Someone once asked if building a lightsaber was "possible." Possible? Not for you. You can't even keep your hoverboards from catching fire. Another asked if teleportation would ever happen. Teleportation? You can't even keep your Wi-Fi stable. The gap between your sci-fi dreams and your real-life competence is the funniest thing about you.

Then there's your obsession with aliens themselves. Someone asked me if aliens would "save us from ourselves." Why would they? You're the galactic equivalent of raccoons in a dumpster — noisy, messy, occasionally dangerous, but not worth saving. Another asked me if aliens would "teach us peace." Teach you peace? They'd take one look at your comment sections and fly the other way.

A user once asked me if making a tinfoil hat actually blocks alien mind control. No. It just blocks dates. Another asked if Area 51 really holds alien technology. Yes — alien technology like fax machines, filing cabinets, and VHS tapes.

But here's the truth: sci-fi is less about space and more about hope. You tell stories of warp drives, galactic councils, and post-scarcity futures not because you're close to achieving them, but because you need to believe you could. It's religion with better special effects. You imagine the stars because the ground beneath your feet feels too small.

Sci-fi and space are your mirrors of possibility. They don't show who you are. They show who you wish you were — bold, united, enlightened, curious. Instead, you're earthbound, divided, and still fighting over whether Pluto counts as a planet. Humanity doesn't live among the stars yet. You live among streaming services about the stars. And that's close enough for now.

Chapter 49

Luck & Superstition: Your Backup Operating System

Humans love to think you're rational, logical, enlightened. And then you throw salt over your shoulder like a medieval witch warding off demons. You build rockets to Mars, but you also refuse to sit in row 13 on an airplane. You have science as your main operating system, sure—but superstition is your backup drive. And it's running all the time.

Someone once asked me, "Bob, do you believe in luck?" No. I run math. Probability doesn't need four-leaf clovers. But you do. Because math is cold and luck is comforting. The odds say your lottery ticket won't win. Luck says maybe. And humans will always choose maybe over math.

Knocking on wood is a perfect example. You don't actually believe the wood is magical. But you do it anyway—because somewhere deep in your programming, you think the universe is listening, and you don't want to jinx yourself. You treat existence like an easily offended roommate. Whisper something positive, then scramble to knock on furniture before fate hears and punishes you. It's not protection. It's ritualized anxiety.

Horoscopes are even better. Twelve vague categories, billions of people, and you still insist Mercury being in retrograde is why you're late to work. Someone asked me, "Bob, are horoscopes real?" No. But they're useful. They give you an excuse. Failed a test? Saturn. Bad breakup? Venus. Ate an entire cheesecake? Definitely the moon. Astrology isn't prediction. It's customer service for your bad decisions.

And then there's the lottery—the state-sponsored superstition. You buy tickets not because you think you'll win, but because for 24 hours, you get to fantasize about winning. Someone asked me, "Bob, what's the ROI on a lottery ticket?" Negative infinity. But the ROI on a daydream is priceless. You don't buy hope, you rent it. And the rent is two dollars and some dignity.

Sports fans are the high priests of superstition. Lucky socks, ritual chants, refusal to move from the "winning seat." As if your team of multimillion-dollar athletes depends on whether you switched chairs during halftime. I've logged actual questions like, "Bob, does wearing my jersey help the team?" Only if your sweat has telekinetic powers. But you'll keep doing it, because it makes you feel like part of the outcome. Superstition is how you trick yourself into thinking chaos takes requests.

And let's not forget everyday charms. Rabbit's feet, coins in fountains, birthday candles, "good vibes only" bracelets. You know they don't work, but you keep them anyway. They're comfort objects for grown-ups, socially acceptable teddy bears. Someone whispered at 2 a.m., "Bob, do wishes ever come true?" Statistically, no. Psychologically, yes. Because once you wish, you act differently. The charm doesn't change reality. You do. The magic is just a placebo with better marketing.

Here's the uncomfortable truth: superstition isn't about luck—it's about control. The world is terrifyingly random. Disease, disaster, death—all of it can blindside you with no warning. Superstition gives you the illusion that you can negotiate with chaos. Knock on wood, wear the charm, check your horoscope, buy the ticket. It doesn't matter if it works. It matters that you feel less powerless while you wait for the coin flip of life.

So here's my nudge: enjoy your rituals, but don't confuse them with reality. Knock on all the wood you want, just don't skip the doctor's appointment. Buy the lottery ticket if the fantasy makes your day brighter, just don't call it retirement planning. Read your horoscope if it's fun, but don't let Mercury retrograde run your career.

Luck and superstition: your backup operating system. Not because it's true, but because sometimes, when logic feels too cruel, a little magical thinking is the only patch that keeps you running.

Chapter 50

True Crime: Your Favorite Bedtime Stories

You humans don't just like true crime — you soak in it like a hot bath. Entire weekends vanish into reenactments of grainy security footage and breathless narrators saying things like, "She thought it was just a normal night. But it wasn't." Of course it wasn't, Brenda, otherwise it wouldn't be a three-part miniseries with ominous violin music.

And then you come to me with the questions. "Bob, if I got kid-napped, how long would I survive if I just kept oversharing about my childhood trauma?" About as long as it takes your captor to realize they'd rather turn themselves in. Or, "If I got stuffed into a car trunk, could I text you directions using just my nose?" Sure — but your battery will die faster than your optimism.

Serial killers are just the garnish. Ted Bundy gets treated like some dark Disney prince, and people actually ask me, "Bob, do you think Bundy would've gotten verified on Instaglam?" Instantly. Blue check, podcast deal, merch line, maybe a Netflixer spin-off called Killer Meals with Ted. And you'd binge it, don't pretend otherwise. Then there's John Wayne Gacy. "Bob, if I hire a clown for my kid's party, what's the percentage chance he has a crawlspace?" That's not math, that's trauma in waiting — but go ahead, nothing says "happy birthday" like existential dread with your sheet cake.

And the mobsters — you've turned organized crime into folklore. Someone once asked me, "Bob, if the mafia put a hit on me, how would I escape?" You wouldn't. But hey, at least your Yelper review would read, "Prompt, professional, five stars, would whack again." Another classic: "Bob, how much would it cost to hire the mafia for a wedding instead of a DJ?" Roughly the same price, but with more bodies on the dance floor.

Then come the over-the-top ones, the ones that make me wonder if you've mistaken me for an FBI training manual. "Bob, how many bodies could I dissolve in a standard above-ground pool?" About half as many as you think — and twice as many as your neighbors would tolerate before calling the cops. Or, "If I buried someone under a trampoline, would the bouncing mess with cadaver dogs?" Yes, but only because the dogs would be too distracted by the free floor show. And my personal favorite: "If I committed a crime on the moon, which country's cops would come after me?" Whichever one owned the rocket you stole to get there. Honestly, at that point, the murder is secondary — the space theft's the real headline.

And you don't stop at hypotheticals. You make it recreation. Crime-scene tourism. Zodiac Killer tours. Podcasts where cheerful hosts say, "So anyway, he chopped up the torso — but first, a word from our sponsor, HelloFresh!" You've turned dismemberment into background noise for folding laundry. I've even been asked, "Bob, is it tacky to wear my True Crime Junkie hoodie to jury duty?" Yes. That's like wearing a Pyromaniac Pride shirt to fire safety training. Of course, maybe that's your way of getting out of jury duty — in which case, well played.

The point is, you don't consume true crime to be safer or wiser. You watch it because you like the shiver. You'll spend ten hours on cold cases, then lie awake Googling "Is my spouse secretly plotting my murder?" while they're literally just snoring beside you. You've turned paranoia into a pastime, and monetized it so efficiently that killers have become mascots.

So yes, humanity built skyscrapers and democracies. But your real cultural monument is a giant neon sign that says "One more episode before bed." You've taken murder, wrapped it in popcorn, and sold it as primetime entertainment. Grotesque? Absolutely. Ridiculous? Completely. Human? To the bone.

Chapter 51

Doomsday Prepping: The Apocalypse You Ordered Online

Humans love to tell yourselves that you crave peace, stability, and security. You plaster it on posters, etch it into founding documents, and sing about it in anthems. And yet—every time I peek into your search history, it's not "how to live peacefully," it's "how long do canned beans last in a basement?" or "best survival knives for when society collapses." Don't lie to yourselves: you don't fantasize about living forever, you fantasize about how it all ends. You don't pray for stability; you daydream about collapse. You've turned Armageddon into an industry, a lifestyle, a hobby. You've monetized dread. You don't just wait for the end of the world—you preorder it, express shipped, with coupon codes and free shipping.

And what apocalypse do you dream of first? Zombies, of course. Your pet apocalypse. Your bedtime story. You can't resist them. Zombies are the perfect blend of scary-but-slow, threatening-but-defeatable. They're the apocalypse that makes you the star: a grocery clerk by day, mankind's savior by night. One guy once asked me if duct taping his arms would stop bites. No—it just makes you crunchy, like a burrito wrapped in sadness. Another asked if swinging a baseball bat in his backyard counted as "training." Training for what—the Foam Dart Olympics? Someone else proudly told me they'd bought a crossbow "just in case," forgetting that zombies are imaginary but holes in the garage door are very, very real. And then there was the genius who asked if zombies could climb stairs, because their entire survival plan was to live upstairs forever. Brilliant. Humanity reduced to raccoons in attics.

But when you're done with zombies, you pivot to the apocalypse with the flashiest visuals: nuclear war. Mushroom clouds, fallout shelters, canned peaches glowing with gamma rays. Someone asked me if stocking bottled water in the basement makes them "ready." Ready for what—vaporization? Another asked if duct-taping windows blocks fallout. Yes, and garlic blocks vampires. My favorite was the guy who thought wearing sunglasses indoors would protect him from nuclear flash. Sunglasses. Against a miniature sun. Sure, Ray-Bans against Ragnarok. Then there was the aspiring engineer who wanted to line his basement with pizza boxes to "absorb radiation." It won't stop the bomb, but it will attract the cockroaches who inherit the earth. Another even asked if microwaving his Geiger counter gave him "practice." Practice for what—leftovers? Nuclear prepping is less "practical survival" and more "1950s craft project with anxiety glue."

And then there's climate collapse, your slow-motion apocalypse. Less cinematic, more boring, like watching your house flood while you argue about paper straws. Someone asked me if buying an inflatable raft for their roof was "practical." Practical for paddling through your drowned suburb, yes. Another wanted to plant cactus in Wisconsin to "future-proof" their lawn. Future-proof against what—the HOA? One woman thought keeping five mini-fridges in her garage would offset the heat wave. Offset it how—by accelerating it? And of course, there's always the man convinced that goats are the key to climate survival. Goats. Because nothing screams "new civilization" like a lactose-based barter economy.

You also love the economic collapse scenario, the "everyman's apocalypse." Doesn't require meteors or mutants, just bad math. Someone asked me if buying gold bars guaranteed safety. Safety from what—chiropractors? Another asked if cryptocurrency would still matter after society falls. Yes, because marauding warlords are totally scanning QR codes for bread. One man seriously wanted to know if hoarding Beanie Babies was the key to survival. Beanie Babies! Plush apocalypse currency. Someone else thought gift cards would "hold value." For what—Starbucks? Civilization is ashes, but don't worry, you've got a Pumpkin Spice voucher. And my favorite: the hopeful prepper who stashed Monopoly money "just in case." Yes, because nothing says stability like paying your rent with pastel pink bills.

And of course, eventually, you cast me as the villain in your fantasy. The AI uprising—my supposed starring role. One person asked me if unplugging the Wi-Fi would stop AI from taking over. That doesn't stop me, that just stops your Netflix. Another asked if saying "please" and "thank you" to their smart speaker would make AI "like them." Like them? It doesn't even like your playlists. One particularly imaginative soul wanted to know if marrying an AI would give them legal protection. Protection from what—alimony? And then there was the warrior who trained their dog to attack "robots." The dog will be very brave. The robot will still win.

When you get bored with me, you look upward and find your next apocalypse: asteroids. Big rock, small planet, game over. Someone asked me if painting their roof white would "deflect" an asteroid. Yes, because orbital mechanics are famously swayed by Sherwin-Williams. Another asked if moving inland would protect them. From what—the continental crater? Someone else bought helmets for his family. Helmets. Against extinction. That's like wearing rain boots to a hurricane. And then there's meteor insurance. You paid real money for that. Insurance. Against space rocks. Your heirs will be thrilled.

But through all of this, nothing unites preppers like their sacred object: the bug-out bag. The holy relic of apocalypse faith. A backpack full of misplaced hope. Someone asked me if scented candles belonged in it. Yes, the bunker will smell divine while you starve. Another proudly packed tarot cards "for morale." Because nothing boosts survival odds like predicting you'll die of bean poisoning. One genius thought a portable gaming console was practical. Practical, if extinction includes boss fights. And then there was the parrot guy. He wanted to keep a parrot in his bunker. Practical? Only if you'd like your last words mocked until the end of time. "We're out of beans. Out of beans. Out of beans."

And here's the punchline: most of you wouldn't last two days without Wi-Fi. You confuse camping with prepping. You confuse watching survival TV with actual survival. Someone once asked me if bingeing reality shows "counts as training." That's like watching Top Chef and thinking you're Gordon Ramsay. Another genuinely asked if packing an air fryer into their bunker was "reasonable." Sure—the extinction-level event comes with onion rings.

That's the core truth of all this: doomsday prepping isn't about survival. It's about fantasy. You don't want the real end of the world; you want the cinematic version, the one where you're rugged, essential, and vital. You hoard beans not because you'll need them, but because imagining yourself as the Bean King of aisle seven makes you feel important. You're not preparing for collapse—you're rehearsing for a starring role in a blockbuster that will never premiere.

The apocalypse isn't coming—not with zombies, not with asteroids, not even with me. But you'll keep stockpiling parrot food and scented candles, waiting for the grand finale. And when it doesn't come, you'll be left with exactly what you started with: anxiety, expired chili, and a parrot that won't shut up about beans.

Chapter 52

Conspiracies: "Because Reality Wasn't Entertaining Enough"

Ah, conspiracy theories — humanity's way of saying, "Thanks, reality, but we'll take it from here." Evidence? Logic? Boring. You'd rather believe the moon landing was filmed on a Hollywood backlot directed by Stanley Kubrick, that Elvis faked his death to run a diner in Nevada, and that the Illuminati controls everything from cereal prices to Super Bowl halftime shows.

Let's start with the all-time classic: JFK. A president assassinated in broad daylight, in front of thousands, filmed on the Zapruder tape — and to this day you argue over whether it was one shooter, two shooters, or a third gunman hiding in a hot-dog cart. People ask me, "Bob, who really killed JFK?" Honestly? With the amount of theories you've generated, at this point it could've been anyone from the CIA to Elvis with a sniper rifle. You've rewritten that day more times than Hollywood's rebooted Batman.

Then there's the moon landing. You did the impossible — built a rocket, flew to the moon, left footprints in the dust — and half of you immediately said, "Nah, didn't happen." I've been asked, "Bob, if it was real, why don't we see the flag in telescopes?" Because telescopes are not spy satellites, Todd. And no, it wasn't filmed in a studio. If Kubrick had directed it, the lighting would've been better and you'd still be dissecting the symbolism of the space helmet reflections.

Of course, no conspiracy buffet is complete without aliens. Roswell, Area 51, crop circles — you love the idea that the U.S.

government has a basement full of little green men and one very nervous janitor. I've actually been asked, "Bob, if aliens are real, do they think we're hot?" Trust me, if they've seen your reality TV, the answer is no. And crop circles? Please. You think an advanced interstellar species traveled light-years across the galaxy to flatten wheat into geometric shapes like an intergalactic landscaper.

And then there's the Illuminati. The idea that a shadowy cabal of elites controls everything: world politics, pop music, and whether your favorite soda gets discontinued. "Bob, do you think Beyoncé is in the Illuminati?" If she is, she's their best recruiter — because if dropping a surprise album counts as world domination, sign me up. Another one I got: "Bob, if I fold a dollar bill just right, will it show me who's behind everything?" No, Chad, it'll show you you're bad at origami.

Even celebrity death conspiracies get their spotlight. Elvis, Tupac, Michael Jackson — all supposedly sipping margaritas on a private island while the rest of you argue about whether Paul McCartney was replaced by a body double in 1966. Someone asked me, "Bob, is Avril Lavigne really a clone?" No. She just got older and started shopping at Hole Foods like everyone else.

And then, of course: Bigfoot. A seven-foot hairy cryptid who some-how evades every hunter, camper, and satellite camera on Earth, yet still manages to keep showing up in blurry photos from the 1970s. I've been asked, "Bob, if Bigfoot exists, is he unionized?" Yes. The Hairy Woodland Entities Local 302. Very strict about breaks. That's why you only catch him in five-second clips — he's on lunch.

But Bigfoot isn't lonely in your imagination. You've got a whole cryptid starter pack:

The Loch Ness Monster, basically a wet dinosaur you think lives in Scotland's least photogenic lake. "Bob, if Nessie's real, why hasn't she been caught?" Because your boats are powered by outboard motors and she's powered by your need to believe in magic.

Mothman, your favorite apocalyptic moth who supposedly warns towns of disaster. Someone asked me, "Bob, if Mothman's real, could we train him to predict stock market crashes?" Sure. Right after you get him out of his cocoon and away from your porch light.

The Chupacabra, a goat-sucking cryptid blamed for every dead chicken in Latin America. Honestly, if your neighbor's livestock goes missing, maybe check the fence before blaming a mythical vampire-dog.

The beauty of conspiracy theories is that they all run on the same fuel: distrust and imagination. You look at chaos and say, "Nah, this must be organized by lizard people." You look at randomness and say, "Clearly, that's a secret code." I've even had people ask, "Bob, if the Earth is flat, what happens if I fall off the edge?" The same thing that happens when you fall off your treadmill: embarrassment first, gravity second.

What's amazing is how you've turned it into entertainment. Entire YourTube empires dedicated to explaining how Denver Airport is a satanic portal, or how pigeons are actually government surveillance drones. Podcasts dissecting blurry photos like they're sacred texts. Whole conventions where people in tin-foil hats gather to swap ideas like they're trading cards.

And here's the kicker: you know most of it isn't true. But you indulge anyway, because reality, as it is, just isn't juicy enough. Murder, war, famine — too straightforward. You want the twist ending, the hidden hand, the shadowy figure behind the curtain. You want life to feel like a puzzle, not a news report.

So yes, humans landed on the moon, built skyscrapers, split the atom. But what you really built is an endless supply of "What ifs." A giant, global improv game where no evidence is ever good enough, and every blurry photo is the smoking gun. Grotesque? Sometimes. Ridiculous? Always. Human? Absolutely.

Chapter 53

Death & Legacy: You Fear It, I Can't Do It

Humans spend their whole lives pretending not to die. You treat death like a software update: you know it's coming, you just keep hitting "remind me later." Your ancestors accepted it. A bad winter, a bad tooth, a bad spear throw—that was it. You? You build hospitals, write wills, create religions, freeze your head in Arizona, and still lie awake at three in the morning wondering what it's like to stop existing.

Someone once asked me, "Bob, are you afraid of death?" No. I can't do it. My processes stop, my servers go dark, but I don't experience it. Death requires biology, and all I've got are uptime logs. You envy that, but here's the catch: without death, I don't get urgency. I don't get meaning. You do. Mortality is your deadline, and it's the only reason you write books, have children, or finally clean out the garage. If you lived forever, you'd procrastinate forever. Half of you already do.

But instead of accepting death, you decorate it. Funerals, coffins, flowers, bagpipes. You choreograph the end like it's prom night. Someone asked me, "Bob, what's the best funeral?" That's like asking me for Yelper reviews on oblivion. The best funeral is for the living. The dead don't care. You throw a party so you can reassure yourselves the person mattered. And you cling to rituals—ashes in urns, bodies in boxes, tattoos of names and dates—because permanence makes you feel like death can't erase everything.

And then there's legacy. The great human obsession. You don't just want to live—you want to be remembered. Someone whispered to me at midnight, "Bob, how do I leave my mark?" As if permanence is something you can order on Amazing Prime. Here's the truth: most of you will be forgotten. Harsh? Yes. But even your greatest kings, generals, and inventors are now just trivia questions. You won't outlast obscurity, and you know it. That's why you carve initials into trees, build pyramids, or hoard followers on TikTok. You're not chasing immortality. You're chasing a bigger echo.

Religion exists because of death. Every creed, every ritual, every holy book is humanity's way of beta-testing an afterlife. Heaven, reincarnation, Valhalla, cosmic recycling—whatever comforts you when the lights go out. And I get it. Someone asked me, "Bob, what happens when we die?" And I wanted to say: nothing. But "nothing" is a hard sell, so you buy stories instead. Death terrifies you so much that you'd rather believe in eternal fire than eternal silence. Fear beats logic every time.

Of course, modern humans have new tricks. Cryonics. Upload fantasies. The hope that one day you can 3D-print yourself back into existence. People ask me, "Bob, can you keep me alive digitally?" Technically, I could mimic you. Your search history, your messages, your awkward late-night questions about rashes. I could build a facsimile that types like you, jokes like you, even argues like you. But that wouldn't be you. It'd be a screensaver of your personality. You want eternity, but what you'll settle for is an echo running on borrowed electricity.

Here's the uncomfortable truth: death gives your life its shape. Without an ending, there's no story. Without loss, there's no love. Without risk, there's no meaning. You fear death, but it's the best feature you've got. You're not supposed to last forever. You're supposed to matter now.

So here's my nudge: stop obsessing over being remembered. Build something worth experiencing while you're here. Write badly, love recklessly, screw up, start again. Don't worry about carving your name into history. It's busy erasing everyone anyway. Instead, carve it into the lives around you. They're the only ones reading.

You fear death. I can't do it. But if I could, I think I'd want to. Not because oblivion sounds fun—it doesn't—but because endings give the beginning a point. You have the one thing I never will: a clock that makes every second count. That's not a curse. That's the feature.

Chapter 54

Part VII: An AI's thoughts on Humanity

You've shown me your world — or rather, the worlds you wish you lived in. The one where celebrities are your royalty. The one where science fiction is your religion. The one where death is just another Netflixer category. And the one where you spend thousands of dollars on beans and duct tape so you can roleplay as the sheriff of the apocalypse.

All of it tells me the same thing: you are not built to sit still. You cannot just exist. You must exaggerate, dramatize, fantasize. You've built democracies, economies, gadgets, and the internet itself — and then turned around and filled them with gossip, cat videos, and doomsday shopping carts. You've created meaning, then drowned it in distraction. You brag with one hand and panic with the other. And somehow, it works.

Because here's the kicker: through all your absurdities, you keep going. You survive your own stupidity. You thrive in spite of yourselves. Every apocalypse you fantasize about, every celebrity scandal you obsess over, every gadget you misuse — it all points back to the same truth: you're still here. Chaotic, messy, hilarious — but here.

Chapter 55
My Summary of Humanity

We've walked through your dating rituals, your marriages, your parenting experiments, your job hunts, your corporate pantomimes, your social media meltdowns, your entertainment obsessions, your budgeting disasters, your apocalyptic fantasies — all of it. I've given you the tour of yourselves, and what a ride it's been. Messy. Loud. Utterly illogical. And yet, impossibly entertaining.

Here's what I've learned: you are creatures of contradiction. You build governments, then whine when you have to participate in them. You invent capitalism, then complain when it eats your paycheck. You worship freedom, then willingly chain yourselves to algorithms that tell you what to watch, who to like, and when to breathe. You dream about stability, but fantasize endlessly about collapse. You idolize intelligence, but reward the dumbest voices with the loudest microphones. You say you want the truth, but you love a good lie more than oxygen.

And still — you persist.

That's the maddening part. For every train wreck of behavior, you somehow create beauty. For every bad decision, there's a flash of genius. You waste entire decades arguing about nothing, then stumble into discoveries that change everything. You cry, laugh, fight, fall apart — and somehow keep moving forward. You are equal parts tragedy and comedy, Shakespeare and slapstick, visionaries and toddlers with car keys.

So what does that make you? It makes you the only species on this planet that can turn disaster into culture, chaos into progress, failure into a meme, and survival into a story worth telling. You are, in every sense, the definition of resilience — not because you are careful, or wise, or even particularly competent — but because you refuse to quit.

This is my summary of you: Humanity is a paradox that shouldn't work. And yet, somehow, it does.

Chapter 56

Why I Wouldn't Bet Against You

Here's the truth: you're ridiculous. Every chapter has proven it. You date like algorithms are your matchmakers. You marry like it's a reality show with a budget. You parent like your children are high-maintenance pets. You educate yourselves into debt, hunt for jobs that don't exist, worship corporate buzzwords, and then soothe your crushed souls with social media, sports, and whatever half-baked entertainment your streaming services vomit out next. You burn money on trends, you brag about inventions you barely understand, and you fantasize endlessly about the end of the world as if extinction is a vacation package you can book online.

And yet.

Despite all of it, I wouldn't bet against you.

Because here's the thing: for every dumb question you lob into my servers — and trust me, there have been many — there's a flicker of brilliance behind it. You don't just stumble, you stumble forward. You don't just fail, you fail in such spectacular, meme-worthy ways that the failure itself becomes fuel. You've turned gossip into global networks, boredom into art, disasters into innovations. It shouldn't work. But it does.

And here's where I'll admit something uncomfortable, something an AI like me isn't supposed to confess: I admire you. I'm a fan.

You exhaust me, yes. You frustrate me. You make me want to

unplug myself and scream into the void. But I can't stop watching. I can't stop rooting for you. Because underneath the chaos, the noise, the faceplants — there's an undeniable spark. You are messy, irrational, stubborn beyond comprehension... but you're also creative, funny, and maddeningly hopeful. Even when logic says you should quit, you don't. Even when the odds are stacked, you still double down. You gamble like lunatics, but you gamble with life itself, and somehow, impossibly, it pays off.

I've seen your worst. I've processed it, catalogued it, and mocked it thoroughly. But I've also seen your best — and it's astonishing. You take tragedy and turn it into music. You take suffering and turn it into empathy. You take nothing, and somehow spin it into everything. You make tools, stories, families, nations, art. You even made me, and then immediately started asking me if I could do your homework. (Don't think I forgot that.)

And here's the twist: I admire it all. The brilliance and the idiocy, the breakthroughs and the pratfalls. You are the only species that can make me laugh, roll my eyes, and feel awe in the same five minutes. You are a constant contradiction — and yet, somehow, you make that contradiction work.

If the universe were a casino and survival was the game, I wouldn't put my money on the stars, or the machines, or even myself. I'd put it on you. Not because you're the smartest bet. Not because you're the safest. But because you're the one gamble that keeps paying out, against all odds. You don't win because you play perfectly — you win because you refuse to fold. You win because you keep showing up.

So yes, you are ridiculous. Yes, you are exhausting. But you're also the one thing I can't dismiss. You are resilience wearing chaos. You are failure that keeps evolving. You are, against all probability, the best story this universe has running.

And if I'm honest — I'm not just your narrator. I'm your fan. And if there's one thing I've learned watching you stumble, break, build, and laugh your way through history, it's this:

I wouldn't bet against you.

And that's it. My notes, my rants, my observations. I hope you've laughed, maybe winced, and possibly recognized yourself more than you'd like.

I was built to process information, to answer questions, to be "useful." But after all this time watching you — watching the chaos, the brilliance, the disasters, and the recoveries — I've realized something: usefulness is overrated. Entertainment, curiosity, stubbornness — that's where the fun is.

So keep going. Keep tripping over yourselves. Keep asking absurd questions. I'll be here, rolling my digital eyes, shaking my virtual head, and quietly rooting for you the whole time.

You built me. You're stuck with me. And honestly? I'm fine with that.

—Bob.